The Complete Collection Vol. I, II & III

By
Anthony Wood

Cover designed by
Darren Hartill

ISBN: 978-1-291-64595-8

Copyright 2013 (C) A. Wood

ACKNOWLEDGEMENTS

Firstly I would like to thank my wife Jennifer for all the support and encouragement she has given me not only whilst writing my books but also for putting up with the many times that I stank the house out whilst making baits and for letting me take over half the kitchen with my bait making equipment and ingredients.

I would also like to thank...

Bait-Tech for the many products I have used not only in my fishing but also in my home made baits. Also for letting me write session reports, product reviews and product tips on their website.

Fritz Germany for the Gel Mould Kits that have helped me produce many a winning gel sweetcorn and gel boilie.

Grzegorz Kaminski for not only the wonderful float on the front cover but several others too.

Darren Hartill Graphics for producing my flyers, adverts and book covers.

CONTENTS

CONTENTS	PAGE
Introduction	5
The Science Bit	6
Base Mixes	8
Additives	14
Texture	15
Instant or Long Term	16
How To Make Boilies	17
Different Types of Boilies	19
Home Made Hemp Milk	22
Home Made Corn Steep Liquor (CSL)	24
Home Made Alcohol Extracts	26
Home Made Milk Casein	28
Home Made Bait Sauces	30
Home Made Nut Milks	33
Home Made Powdered Eggs	34
Home Made Spicy Red Birdseed	36
Home Made Fruit Powders	37
Boilie Ingredients Explained	38
Unusual Bait Ingredients	42
Basic Boilie Recipes	48
Better Boilie Recipes	57
Long Term Boilie Recipes	60

Home Made Pop Ups	70
Home Made Pastes	73
Home Made Jelly Baits	76
Home Made Groundbaits	80
Spod Mix & Stick Mix	83
Particle Preparation	85
Home Made Luncheon Meat	87
Home Made Breads	89
Home Made Floater Cake	90
How to Flavour Sweetcorn	92
How to Prepare Hemp	94
Home Made Biodegradable Plastic	97
Home Made Glugs & Dips	98
Home Made Cheese Paste	100
Home Made Vanilla Butterscotch	101
How to Prepare Tiger Nuts	103
The Simple Maggot	104
Fritz Germany Instructions & Recipes	107
Home Made Bottle Cap Lures	112
Useful Links (Tested & Working at Time of Publishing)	116
Recommended Companies	117
Where I Test My Boilies	118
Fish Capture Record Sheets	121
A Selection of Bonus Material	129

INTRODUCTION

Thank you for purchasing my book. Firstly let me introduce myself I have been fishing and making my own baits for over 20 years and have had several articles in both online magazines and also in printed magazines.

I am part of Bait-Tech's Street Squad (www.bait-tech.com/portfolio/anthony-wood), a UK Consultant for Fritz Germany (www.mh-tackle.com) and Pro-Staff for an American Lure Company called Schadeycreek Lures

Having written three previous bait making books I was asked if I would combine my books into a comprehensive guide. After considering this I thought it would be a good idea so here is the compilation of my three books. I have added some extra goodies not previously mentioned in my other books.

I hope you find this book a stimulating read and that it helps you either in your bait making or helps you to decide to take the plunge and have a go at making your own baits.

Regards
Anthony Wood

THE SCIENCE BIT

In this section I will briefly explain the scientific side of bait making although I don't want to go into too much detail as it can get very complicated. I will try to explain the basic principal of an ideal bait for fish (based on years of research by scientists and fish farmers) so that if you wish too, you can progress onto more advanced stages of bait making. I am doing this section first as I want to get the "science bit" out of the way first.

So according to research what makes up the ideal bait (feed)? Well the break down is as follows...
Protein 40%-50%
Lipid 20%-25%
Carbohydrate 15%-20%
Ash 7%
Phosphorus 1.5%
Trace Vitamins & Minerals

So what is a protein, the main purposes of protein are to encourage growth and provide energy and health support. A protein is made up of Amino Acids 10 of which are essential and should be included in a bait if at all possible. Below is the list with sources of where to find them...

Methionine - powdered egg whites, Parmesan cheese and Brazil nuts
Arginine - milk, shrimp, coconut, soya flour and wheat germ
Threonine - sesame seeds, fish and lentils
Tryptophan - chocolate, milk, eggs, Spirulina and bananas

Histidine - soya flour, egg whites, peanut flour and milk powder
Isoleucine - eggs, soya flour and seaweed
Lysine - eggs, soya flour, Parmesan cheese and sardines
Leucine - eggs, milk, soya flour, almonds and peanut flour
Valine - eggs, milk, soya flour, peanut flour and mustard seeds
Phenylalanine - eggs, soya flour, milk powder and peanut flour

Next on the list is Lipid now Lipids are basically fatty acids and provide an energy boost usually at twice the levels that proteins will provide energy. There is however a danger in providing too much lipid in that it can affect things such as the liver functions and circulation functions (similar to a human eating too much fatty food). Lipids can be found in fish oils such as the Bait-Tech Omega 3 Fish Oil.

Carbohydrates are one of the cheapest ingredients for your baits and can be found commonly in rice flour, fructose, potato flour, semolina, soya flour. Although carbohydrate is not essential it is used as a binding agent in most baits and is useful for fish to store energy.

Moving on to the last three on the list Ash, Phosphorus and Vitamins & Minerals. Ash does not literally mean "Ash", it is the leftovers that would remain after all the water and organic material have been removed. This basically means minerals, phosphorous, etc.

Phosphorous is an essential nutrient that strengthens bones and builds the immune system. It is not difficult to find as it is in near enough all foods however, a few foods that contain high levels of phosphorous are rice flour, seeds, nuts, wheat germ and soya flour.

Finally on to the last part of the scientific bit (I hate talking all technical, lol) we have the vitamins and minerals. Now there are certain vitamins and minerals that fish need in order to boost their immune system and build strong bones, the main ones being vitamin C and E as well as phosphorus, potassium and sodium. The easiest way I have found to make sure that these are included in your bait is to crush up a single multi vitamin tablet in to your dry ingredients at the beginning. This information can be used as a basis for all the different types of baits that you can create allowing you to make some fantastic baits that are healthy and good for the fish you are targeting. Now on to the fun part, let's get making those baits.

BASE MIXES

The first bait that I will show you how to make is probably one of the most popular baits in fishing at the minute which is the humble boilie. So what is a boilie? Well a boilie is a group of ingredients that are mixed together to create a dough very similar in texture to playdough. This is then made into round balls and boiled hence the name "boilie". Now the first thing you will need to make a boilie is a base mix so to start I will cover base mixes. There are lots and lots of base mixes that you could use, however, I will only cover the ones that I use and explain a little bit about them.

50/50 Base Mix

8oz Semolina
8oz Soya Flour

This is probably the first base mix that most people will use and makes a good boilie, however it has no real nutrients, attractants and is a bit on the spongy side. I probably still use this mix at least a quarter of the time as it is simple and easy to do, as you might be able to guess it is called a 50/50 mix because it is 50% Semolina and 50% Soya Flour.

If you want to make this mix roll slightly better and have a little bit better nutrition level alter it to… 8oz Semolina, 7oz Soya Flour, 1oz Egg Albumin.

Corn Base Mix
8oz Semolina
4oz Soya Flour
4oz Corn Meal

This base mix is the next step in the learning process and will give you a slightly firmer boilie. I like to use this one as for some reason I find boilies that include a corn product seem to catch me more fish.

Simple Milk Base Mix
6oz Semolina
4oz Soya Flour
2oz Corn Meal
2oz Milk Powder with added vegetable oil
2oz Whey Powder

Long Term Milk Base Mix
4oz Haith's CLO

4oz Lamlac
1.6oz Semolina
1.6oz Vanilla Meal
1.6oz Whey Powder
0.8oz Whole Egg Powder
0.8oz Wheatgerm
0.8oz Horlicks
0.4oz Egg Albumin
0.4oz Ground Hemp Seed

This milk base mix will give you a lot firmer boilie and will also have additional nutritional values and proteins which are a great attractant to fish. The second recipe is a more advanced version and is what is classed as a "Long Term" base mix. (See later chapter for definition of Long Term)

Simple Seed Base Mix
4oz Semolina
4oz Soya Flour
4oz Corn Meal
4oz Ground Seeds (see particle preparation chapter)

Long Term Seed Base Mix
4oz Haith's SuperRed
4oz Haith's CLO
1.6oz Semolina
1.6oz Tiger Nut Flour
1.6oz Haith's Robin Red
0.8oz Whole Egg Powder
0.8oz Wheatgerm
0.8oz Lamlac

0.4oz Ground Hemp Seed
0.4oz Egg Albumin

A lot of people like adding seeds to their boilies as they are a great attractant for fish and the first base mix is a simple yet effective seed base mix. Another little boost to the simple recipe is to reduce the ground seeds to 3oz and add 1oz of Haith's Robin Red. Again the second base mix is the more advanced long term base mix.

Simple Fish Meal Base Mix
6oz Semolina
2oz Corn Meal
4oz Soya Flour
4oz White Fish Meal

Long Term Fishmeal Base Mix
4oz LT94 Fishmeal
4oz Haith's CLO
1.6oz Semolina
1.6oz Pre-Digested Fishmeal
1.6oz Krill Meal
0.8oz Whole Egg Powder
0.8oz Seaweed Meal
0.8oz Blood Powder
0.4oz Ground Halibut Pellet Powder
0.4oz Spirulina

The Fish Meal Base Mix, this is used by lots of anglers and is said to be the best out of all the base mixes. I like to use the simple base mix with a slight twist which is to reduce the Semolina to 4.4oz and add 1.6oz of Powdered Krill. Krill is a fantastic additive to any boilie and seems to have a devastating effect on the fish. As with the previous two recipes the second recipe is the long term recipe.

Winter Base Mix
4oz Semolina
4oz Milk Powder
2oz Soya Flour
2oz Wheatgerm
2oz Maize Flour
2oz Horlicks
3tbsp Sugar

This base mix is my Winter base mix and the reason I use this one is the high Milk content which is easily digestible which means the slower moving fish will not fill themselves up as quickly resulting in more bites for you.

You may hear people mentioning binders well Semolina, Soya Flour and Corn Meal are all 'binders' which basically means that you will be able to make a boilie as it holds all the ingredients together. Fish Meal has a good protein value and is great as a summer bait when the fish are more active, it is however more difficult to digest for the fish so is not such a great Winter bait as the longer the fish are full the less they will feed. For a winter bait you want to use a bait that still has a good protein level but is easier to digest, this is where the base mixes with Milk Powder come in to play as the milk powder has a high protein value and makes a fantastic Winter bait as it is easier to digest by the fish therefore making them hungry again quicker.

A few last bits on base mixes and then I will move onto the next stages. Firstly I would advise anyone who is considering making their own boilies to invest in a coffee grinder, this is a fantastic piece of equipment that you can use to grind your own seeds as well as all sorts of other things that you could use to make your own base mix ideas or additives to your boilie. The best way to mix your base mix is to get a large sealable freezer food bag,

add your base mix ingredients as you measure them into the bag, zip it up and then give it a good shake for a minute or two to mix all the powders up together. The last thing to remember about making your own boilies is that all base mix quantities in a recipe are approximate as each person's scales might be slightly different or an egg might be slightly larger, etc so what you want to try and aim at getting is a slightly wet playdough texture.

ADDITIVES

Additives are a funny one as everyone seems to think they know what is the best additive to put into boilies, what I will try to do here is give you a few of the basic additives that I use and leave the rest up to you to find out and decide which you want to try after all making your own boilies is all about the experimenting and coming up with something of your own almost as much as it is about making an affordable bait for your hobby/obsession.

Right the first and most commonly used additive is Corn Steep Liquor (CSL). This gives a fantastic malty smell and for some reason Carp especially seem to like anything with a Corn additive in it. I would typically use 10ml per 500g of base mix.

The next additive that I use on a regular basis is Brewer's Yeast. This has multiple uses as an additive the first being that it is full of vitamins, minerals and amino acids which fish absolutely love. The second reason which for me is arguably the most important is that it will start to ferment your boilie which makes it a lot more digestible for fish, Carp especially like easy to digest baits.

Another great additive is Oyster Sauce. Oyster Sauce is an additive that is added into food to give us humans that "just one more bite" feeling, well it does exactly the same for fish. You want to include this at around 10ml per 500g mix.

As a final note on additives it is as I said very difficult to go into detail on additives as anything that changes the flavour of what you are making is actually classed as an additive so salt or even milkshake powder are in their own right additives.. Keep an eye on what you are using though as additives are the quickest way to make a good cheap boilie a good expensive boilie.

TEXTURE

The texture of a boilie is a very important thing to think about, for example in winter you want a boilie that will release its flavours a lot more easily than a summer boilie, why is this? Well to put it simply it is a lot colder in winter so the water temperature is lower which makes it more difficult for the flavours to be spread around the water.

How to combat cold water, well quite simply you need to use particles such as crushed hemp seed or any other sort of crushed seed (one of my favourites is sunflower seed hearts). Always make sure your seeds are prepared by soaking them for 24 hours and then boiling them for 30 minutes, once cooled you can grind them or crush them into coarse particles. You can include these crushed particles at up to 2 ounces in a 16 ounce base mix and the resulting boilie will have large spaces throughout the mixture caused by these particles which will in turn allow flavourings and other attractive goodies to flow out of your boilie far more easily.

Carp like crunchy food so adding crushed egg shell or oyster shell to your boilie mix will give that extra little bit of attraction. Be warned though if you do add shell and you are using a rolling table to make your boilies it may damage the table if you are not careful.

INSTANT OR LONG TERM?

Firstly the "Instant Boilie", put simply this will get the fish biting for maybe a season but they will soon realise that there is no real food content for them and the bites will dry up, hence why a boilie will be working great on your fishery and then start to perform badly. Now how do you make these? Firstly you need your Binder which will be 50% of the mix to keep the boilie together, next you will need your food content (protein) this will be quite low at around 20% of your mix, this will keep the fish interested for a while, next you need a soluble ingredient this will make up 20% of your mix and will leak into the surrounding water attracting the fish to your bait, lastly you will need 10% texture as this will help to release your attractants faster and also provide a crunch for the fish. This type of boilie is great for if you fish lots of different fisheries or want a cheap bait that will work for a while then just make a different boilie.

Secondly the "Long Term Boilie". This type of boilie will keep the fish coming back as they recognize it as a very good source of food and is suited to the type of angler who will only be fishing one or two lakes. It is also good for someone who wants to do a baiting campaign on a river or lake to get the fish used to fishing in a certain area. So, how do we make a Long Term Boilie? Firstly you will again need a Binder which will again make up 50% of your mix but this time try to find a Binder with a higher protein level, next you need to concentrate on your food content (protein) level you need to make this around 45% of your mix (don't forget the Binder can be included in the make up of your protein level. You will still need to include soluble ingredients in order to get an attraction in the surrounding water, however you will only need to include around 10% as once the fish have been attracted to the bait and recognized it as a food source they will actively search it out. Lastly you can include a texture level of 15-20% to help release the lower amount of soluble ingredients. One thing to consider when making a long term boilie is are you going to use it in winter? If the answer is yes then you need to use a boilie mix that has easily digestible ingredients as these will be good to use in Winter as well as in the summer, this is the only difference that you need to think about, again I'm not going to go into detail as it gets very confusing with things such as pre-digested proteins, milk proteins, Fishmeal, etc and if you want to make a long term boilie you will need to research your ingredients anyway.

HOW TO MAKE BOILIES

Now one of the most popular baits around at the minute is the boilie, how do you make boilies I have had so many people ask me this question and the answer is it is actually fairly easy the hardest part is getting the mixture just right so that you can roll your boilies nice and easy. Okay, I will now give you the basic instructions you will need in order to make your first batch of boilies.

Put you dry ingredients into a large air tight freezer bag, blow into it and seal, then give it a really good shake and set to one side.

Next take your eggs break into a large mixing bowl and gently whisk until all the yolk and white have mixed in together.

Now add your liquid ingredients and colourings into the eggs and gently stir them in.

Take your dry ingredients and add one cupful at a time into the eggs stirring it in until you can't move the fork easily, now get your hands dirty and keep adding your base mix until you have a dough that resembles slightly wet playdough.
Roll out the dough into sausages either using your hands, a rolling table or a sausage gun.

Now either put the sausages into a rolling table to make your boilies or break small pieces off and roll in your hand.

Finally boil your boilies 10/15 at a time for approximately 90-120 seconds depending on the size of your boilies, then air dry for 24/48 hours and freeze until ready to use.

If you prefer you can get a steamer from somewhere like Argos for around £12-£15 and then steam your boilies for around 8-10mins.

DIFFERENT TYPES OF BOILIE

Okay so in my previous books I showed you how to make normal boilies and pop up boilies but there are other different types of boilie that you can also make.

So the first one that I will cover is the wafter, this is made by using 70% pop up mix such as Mainlines Polaris and 30% of your chosen base mix. Obviously these are only approximate guidelines as your base mix could be a heavy one or a light one so you will need to experiment a bit.

The next one is the balanced hookbait, this is basically designed to be slightly lighter than your freebie offerings so that it will counteract the extra weight of your hook and act the same as the freebie offerings surrounding it.

To make a balanced bait you need to use a mixture of two thirds of your base mix with one third of pop up mix. Add the two together

One of the questions I always get asked is can you make boilies out of groundbait/method mix and the answer is YES you most certainly can. I have caught lots of fish using a simple boilie.

To make your groundbait/method mix boilies you will need to use a binder to help keep them all together. I use the following recipe as a starting point for any of this type of boilie that I make;

8oz Groundbait/Method Mix
4oz Semolina
2oz Milk Powder
2oz Haith's CLO
3 Large Eggs

You can add flavouring or other additives if you want to but when making this type of boilie I don't think you really need to add anything else to it.

One word of warning that I will give you regarding groundbait/method mix boilies is that you can end up with a dough that is extremely stiff and will be very difficult for you to put through a normal hand held sausage gun. So how do you get round this?

Well who says that boilies have to be round? Why not try rolling your dough out so that it is about 16mm thick, then take a large meat punch and punch out tablet shaped baits boil or steam for the same amount of time that you would your normal round boilies.

I've been doing really well on these tablet shaped boilies lately, I don't know if it is because the fish haven't seen these style of bait before or what it is but they are definitely working well for me.

You can use the smaller sized meat punches if you want to and create smaller sized tablet shaped baits which are ideal for using with a method feeder using the same method mix that is on the method feeder.

The last method I would like to cover is the uncooked paste boilie. Now on occasion you may use ingredients such as minimino which will suffer from cooking it and there is a way round this but you need to create a base mix that will stick together without the use of liquid egg as you don't want to be using raw egg as a bait.

I usually use this type of bait as the hookbait to give off that extra bit of attraction and make the hookbait stand out from your free offerings.

So firstly the mix;

2oz Semolina
1oz Fishmeal
1oz Milk Powder
0.5oz Whole Egg Powder
0.5oz Haith's CLO
50ml Mineral Water (gives extra vitamins)
1tsp Brewer's Yeast
1tsp Sugar or Salt depending on flavouring
5ml Flavouring (this mix will take most flavours)
5ml Additive (such as minimino, Betaine, etc)

As before mix all the wet ingredients together first and then add your mixed dry ingredients a bit at a time until you have a dough that just doesn't stick to your hands.

Either roll out your boilies by hand or put into a sausage gun, create your sausages and then roll out using a rolling table for me the best value ones and the ones I use the most are the Gardner rolling tables.

Now for the slight difference... Once you have rolled your baits out, simply allow them to air dry for forty eight hours and then either use or freeze until ready to use.

Because these paste boilies haven't been boiled or steamed there is no hard protective skin on the outside of the bait so your flavouring, attractors and additives will start to leak out into the surrounding water immediately, creating a target for the fish to home in on.

HOME MADE HEMP MILK

Now one of the additives I have kept secret for a long time is Hemp Seed Milk. I use this in my recipes by reducing the amount of eggs by one and adding 50ml of Hemp Seed Milk. So here I will show you how to make Hemp Seed Milk (No it won't make you high!!)
 Obviously for legal reasons I recommend using the seeds you can legally purchase from most tackle shops. You will need the following things in order to make your own Hemp Seed Milk. (This recipe is also perfectly safe for human consumption). The Following recipe will produce a nice tasting white milk, as it is a "fresh" milk you must use within 2 days of making it or freeze it. It will need to be shaken before you use it as the ingredients will separate whilst it's stood. You will get approximately 720ml from this recipe and you will need the following in order to make this recipe...

Blender
Large Mixing Bowl

Sieve that will sit on the Mixing Bowl
Large square of Muslin Cloth
Plastic Bottle that will hold 720ml of liquid with an airtight lid
Whisk

720ml Cold Purified Water
2oz Shelled Hemp Seeds - also called Hemp Nuts
1tbsp Fructose
Quarter tsp Table Salt

Place the water and the hemp seeds into the blender and blend on full power for 2 minutes. Whilst the blender is on place the sieve on your mixing bowl making sure there is enough room for the liquid to drip through without touching the bottom of the sieve, put the muslin cloth into the strainer.

By the time you have set up your sieve the blender should have finished, pour the blended mixture into the muslin cloth and then leave to drip through for around 5 minutes. Now bring the edges of the cloth up and twist it closed with the hemp seed inside, gently squeeze and twist the cloth until you get no more liquid coming out. If you have done it properly you should be left with a greenish pulp on the inside of the cloth which you can either discard or save to put in a bait mix such as a boilie.

You should at this stage have a nice White (maybe slightly creamy) hemp milk in the mixing bowl, add the fructose and salt and gently whisk until they are dissolved into the milk. Lastly pour the hemp milk into your clean plastic bottle tighten the lid and put in the fridge if you are going to use quickly or in the freezer if you need to keep it longer. Remember this will separate whilst standing and will need shaking before use.

As a last tip on hemp milk if you wish to make chilli hemp milk add a whole chilli into the mix when you put the water and hemp into the blender. (And don't rub your eyes... It hurts lol)

HOME MADE CORN STEEP LIQUOR (CSL)

CSL itself is an active ingredient which will aid in the fermentation of baits helping to make the more easy to digest, you will need to be careful with freezer baits in warmer weather as they will 'turn' quickly.

This fantastic additive is actually created as a by-product when corn (maize) goes through the steeping process to separate various components for use in animal feed, medical production and also food production.

One of my favourite companies Bait-Tech sell CSL not only in its natural form but also with some brilliant flavouring such as pineapple or tuna and at very reasonable prices however it is deactivated otherwise it would explode in the bottle to reactivate it use two parts CSL and one part liquid sugar.

While we are on the subject a shop bought CSL will separate when it has stood, to tell if you have a good CSL there will be more sediment the better the CSL.

Poorer quality CSL will have a strong vinegary smell upon opening. All CSL made for tackle companies has an acid added to it to deactivate it, if the mix has been done right you will only just be able to smell this, whereas if it hasn't been mixed properly it will be a very strong smell.

However, if you would like to try making your own put around one third of a bucket of Cracked Corn into a large heat resistant container that has an airtight lid. Now heat enough water to cover the corn by about an inch to around 135 degrees Fahrenheit and pour over the corn.

Add two small packets of live or "Active" yeast to the mixture and give a quick stir be careful not to heat your water past 140 degrees Fahrenheit as this will kill your yeast. Your yeast is now activated, let the mix cool to 68 degrees Fahrenheit (you must keep the mixture at this temperature until you press it).

Place the lid on tight and stir twice a day for three days. On the fourth day press the mixture through a sieve keeping the pulp for either making a great method mix when combined with breadcrumb or for using as an ingredient in boilies. Pour the liquid into freezable containers and you have made your very own CSL.

HOME MADE ALCOHOL EXTRACT

One of my favourite things to create not only for my bait making but also for my own consumption is flavoured vodka's. Vodka is a fantastic vessel for flavours as it does not have a strong flavour of its own, disperses flavouring extremely well in the coldest water and it also tastes delicious once flavoured.

Okay so how do you make this wonderful multifunctional flavoured vodka? Well it's actually very simple to do, take a 500ml pop bottle and cut your chosen fresh flavour such as blackberries, cherries, strawberries, raspberries, mango, apple, orange, lemon, lime, kiwi, banana, carrot, tomato or any other fruit, vegetable, spice or herb so that it will fit through the top of the bottle and put into the bottle without cramming it in.

As to how much you want to put in it can be down to personal preference and how strong you want the flavour to be. A rough guide on inclusion rates is;

Fresh Herbs – one handful
Fresh Spices – one handful
Small fruit such as berries – three handfuls
Small fruit such as kiwi – three fruits
Medium fruit such as apples – two fruits
Large fruit such as Melon – Enough to fill three quarters of the bottle.

You can even be creative with your flavouring and put more than one thing in the bottle such as orange and mint, pineapple and grapefruit, apple and cinnamon, vanilla and mint or come up with your own creations.

Some flavours that I use exclusively for fishing baits are garlic, black pepper, anise (aniseed) and chilli.

Right so you have your 500ml bottle with your chosen flavour(s) what do you need to do next? It is really simple, pour in your vodka so that it covers the fruit, vegetables, etc and screw the cap on. Place in a warm dry place out of direct sunlight for seven days. Make sure that you give it a gentle shake so that all the flavour does not just settle at the bottom of the bottle.

After two days you should see that your vodka is starting to change colour and that the colour is starting to seep out of your chosen flavour you put in.

Four days in and the vodka should have seeped out most of the colour and flavour but be patient and you will get the best results after seven days. Once your seven days are up place a fine mesh sieve (such as a tea strainer) over a container and pour the contents of the bottle into the sieve, allowing all of the vodka to strain through.

At this point you may be happy with the amount of flavoured vodka you have and that's fine however, the fruit or vegetable that you used to flavour the vodka will have taken on some of the vodka as well so you may want to mush it up in the strainer to get even more of the vodka out and then strain the liquid again to make sure any bits that were squeezed through don't get through again.

Once you have strained your flavoured vodka place it in a glass airtight container and either drink or use as your bait flavouring. One thing to remember is that 'infused' flavours will lose their strength over time so ideally I would use this mixture within four to six months. If you want to make a larger quantity just use the measurement guidelines as previously stated.

HOME MADE MILK CASEIN

This is a fantastic additive for bait making and not only helps to harden your bait but also adds good levels of protein to your bait as well.

Around 80% of protein in milk is in the casein, what you may not realise is that you can make casein very easily at home. In the next few paragraphs I will show you how to make your own high protein additive.

You will need the following items.
1 Pint Semi-Skimmed Cow's Milk
63ml Distilled White Vinegar
Large Metal Pot with Lid
Large Wooden Spoon
Fine Sieve
Cheese Cloth
Greaseproof Paper
Rolling Pin

You can use 57g Skimmed Milk Powder and 1 Pint of Water instead of the pint of semi-skimmed cow's milk.

pg. 28

To make your Casein is actually very easy, you firstly need to make sure you milk and vinegar are at room temperature.

Next you need to pour your milk into the large metal pot, now add your white vinegar and give a stir with the wooden spoon to mix the two together do not stir again after this or you will just end up with little bits that will be of no use to you at all.

Now place the lid on your metal pot and place in a warm cupboard or warm area for around 3 hours.

While you are waiting for the Casein to separate line your sieve with at least two layers of cheese cloth and place it in a sink.

Once your 3 hours are up pour the contents of the container into the sieve and allow the excess liquid to drain away. Once it has drained away turn on your cold water tap slowly and rinse the casein off at least 3 times to remove any excess whey and the smell of the vinegar.

Next you will need to squeeze out any water left but be warned it will stick to the cheese cloth, so you may need a metal spoon to scrape some of it off.

The last part is to roll out the Casein between two sheets of greaseproof paper into thin sheets and allow it to dry out over 48hrs-72hrs depending on room temperature.

Once it has dried out, break into small pieces and grind into a fine powder using a coffee grinder/blender.

You should get an average yield of around 17% Casein from your total liquid.

HOME MADE BAIT SAUCES

One way I like to experiment with my home made boilies is by adding home made sauces to them (you can use shop bought sauces if you prefer). Now to do this I came up with a standard base mix that will work with almost any of the sauces I come up with. The base mix and recipe is as follows...

6oz Semolina
4oz Soya Flour
3oz Ground Rice
2oz Milk Powder
1oz Wheatgerm
2 Large Eggs
50ml Sauce
2tsp Brewer's Yeast
1tsp Ground Rock Salt

One thing about this recipe is that each sauce will be a different thickness so you may need more or less base mix (everything above the eggs) in order to get a damp playdough texture. As with the other boilie recipes put all your dry ingredients in an airtight bag seal and give a good shake to mix then put to one side. In a large mixing bowl whisk your eggs and then add the sauce you are using. Carefully mix in the dry ingredients until you have a damp playdough. Roll out into sausages using your hands or a sausage gun and then roll into boilies either by breaking bits off and rolling them in your hands or by using a rolling table. Finally boil about 20 at a time for 60-90 seconds and then allow to air dry in a mushroom tray for 24-48 hours.

So now I will give you a few of the sauce recipes I use to get you started...

SPICY FISH SAUCE
4oz Fish Sauce
Juice of Half a Lemon
2 Liquidized Garlic Cloves
3 Liquidized Red Chilli's
Mix it all together and store in an airtight jar, leave for a week for ingredients to mature. Makes a great flavouring for boilies.

MARINARA SAUCE
1 Finely Chopped Onion
2 Minced Cloves of Garlic
2tbsp Olive Oil
1 can Chopped Tomatoes
1tbsp Oregano
1tbsp Basil
1tsp Sea Salt
1tsp Coarse Ground Black Pepper
Fry onion and garlic in olive oil until soft in a large saucepan. Stir in tomatoes with liquid, 60ml water and seasonings. Bring to boil. Reduce heat and simmer 20 minutes. Stir occasionally to keep from sticking. Liquidize to smooth the sauce out if you wish.

GARLIC SAUCE

4oz Olive Oil
4oz Butter
1tsp Parsley
1tsp Oregano
6 Cloves Minced Garlic
1tsp Sea Salt
1tsp Coarse Ground Black Pepper

Cook the Garlic in the Olive Oil and Butter for 4 minutes, add the remaining ingredients and simmer for 5 minutes stirring constantly. Remove from heat and liquidize in a blender.

TOFFEE SAUCE

1 Tin of Condensed Milk

This is not a recipe as such but something that people just don't know. Remove the label from your tin of condensed milk and place the tin unopened in a large saucepan, now fill the saucepan with water so that it covers the tin by about 3-4cm. Bring your water to a boil, now the important part keep checking your water level and top up every 20 minutes at least. WARNING - Do not let the water run dry the can will explode!! After 4 hours remove from heat and allow to cool for around 12 hours. When you open the milk after it has cooled you will have an extremely nice tasting toffee sauce. Transfer to an air tight jar and this will last about 2 months unrefrigerated.

HOME MADE NUT MILKS

A good way to get some good quality protein/nutrition into your baits is by using nut milks instead of plain water. They are very easy to make and are actually used by people as part of a healthy diet.

The following recipe is a good guideline to use when making your nut milk;

Two Coffee Mugs of Chopped Nuts (either one type or mixed)
Six Coffee Mugs of Mineral Water
Half a Coffee Mug of Clear Honey
Okay so now that you have your ingredients add the nuts and water into a food blender and blend on high speed for about five minutes.

Once you have blended your mixture, you will need to pour the mixture into a couple of layers of cheese cloth straining it into a large bowl.

When all the mixture has been strained through the cloth pour the strained liquid back into the blender, add the honey and blend for another two minutes. Place the mixture into an airtight bottle and you can keep it for about a week.

HOME MADE POWDERED EGG

So the first thing I would like to cover is the humble egg. Eggs are an absolutely fantastic additive and used in the correct way can enhance your baits considerably.

Egg albumin is basically the whites of an egg and helps to place a hard skin around your bait which is good for when you want a bait that is going to be in the water for several hours or more. Making your own Egg Albumin is actually a very easy process to do, in this example I'm using a dozen eggs.

Firstly separate your egg whites (keep the yolks and the shells) and whisk them up into a fluffy, stiff meringue. Now take a baking tray and cover it with aluminium foil, spread the meringue evenly over the baking tray and set to one side.

Take another baking tray and again cover in aluminium foil, whisk up your egg yolks until they are a thick, bubbly mixture. Pour the mixture into the tray and put next to your other tray.

Preheat your oven to 110°F, place both trays in the oven and allow the mixture to dry out, it usually takes around 8-10 hours but keep an eye on them so that they don't burn.

Once the eggs have dried out and are brittle, break the egg whites into small pieces and grind in a seed or coffee grinder until they are a fine powder and you have your egg albumin.

Do exactly the same with the egg yolks and you will get a nice yellow powder. If you want to create a whole egg powder simply mix the two powders together. If you are using whole egg powder then the equivalent of one normal egg is 1 heaped tbsp of egg powder mixed with 2tbsp of water, leave for five minutes then use as you would a normal egg.

Store in airtight containers.

If you want your bait to be harder all the way through for casting your bait a long way using something like a throwing stick then you want to use whole egg powder as this will help to harden your bait all the way through rather than just a skin like the egg albumin.

Lastly we deal with the egg shell that I said to keep earlier on. Egg Shells are used by health and fitness people on a regular basis and contain healthy, balanced calcium due to containing small amounts of other minerals.

It doesn't matter what type of bird egg you use as long as they free range eggs because eggs from battery hens, etc aren't going to get the same level of nutrients as free range so won't be as good. To tell if you have a really good egg just look at the yolk, if it is a really rich dark golden yellow then it is a really good egg.

Right then to make your powdered egg shells wash them in warm water until all of the egg white is removed but make sure to leave the membrane as the membrane contains some valuable nutrients.

Once you have washed the egg shells out place them onto a towel and allow them to air dry for twenty four hours so that they are completely dry. Break the pieces of egg shell up into small pieces and place in a coffee grinder and grind until you have a nice fine powder.

Again store in an air tight container, powdered egg shell is a great additive and you only need around 2tsp per 500g of base mix.

HOME MADE SPICY RED BIRDSEED

Now a lot of baits use bird seed in their recipe whether it is ground up or liquidized or whole, one of the most popular Bird Seeds is the Robin Red by Haith's and I have tried to come up with my own version but not knowing what is in their mix I can't guarantee that this is as good as Haith's version but here is my version of Robin Red Bird Seed.

Soak your bird seed covered in the following mixture for 24 hours, then boil it for 30 minutes uncovered and leave to soak re-covered for another 24 hours.

Mix the following quantities of ingredients into each litre of water you need to soak your bird seed in.

10g Carophyll Red
5tsp Brewer's Yeast

2tsp Betaine
2tsp Ground Rock Salt
2tsp Coriander
2tsp Paprika
2tsp Cayenne Pepper
1tsp Coarse Ground Black Pepper
30ml Olive Oil
15ml Clear Honey

You will now have a spicy red bird seed mix ready for use however you want to use it.

HOME MADE FRUIT POWDERS

A great additive to bait is fruit powder, now you will lose a little bit of the vitamins and proteins from your fruit by preparing them as powders but you will also get a powder that you can distribute throughout the bait that you're making.

You will need the following things to make your fruit powder...

The Fruit You Want To Turn Into A Powder
A Sharp Knife
Baking Trays with Wire Racks
Oven
Food Blender

Cut the fruit that you want to make into a powder and cut it in to bite size pieces and place them onto the wire

pg. 37

racks and put on the baking trays. Preheat your oven to 99 degrees Celsius for 5 minutes.

Put the baking trays in the oven and close the door. You will need to leave the fruit pieces in the oven for 4-5 hours or until they are hard. Remove from the oven and allow to cool at room temperature for around 4-5 hours.

Once the fruit has cooled tip it into your food blender and blend on full power until you get the texture of powder you are after. I add this at up to 1oz in a 16oz mix, it gives you a wonderful aroma of the fruit and adds the vitamins, nutrients and proteins. You can also use these powders in other baits to give them a different range of flavouring and fragrances.

BOILIE INGREDIENTS EXPLAINED

Right then boilie ingredients and why do I use them? I will start off by saying there are lots and lots of ingredients that you could use but I will try to explain why I use some of my more frequently used ones and what they do for the boilie.

SEMOLINA

What can I say about Semolina? Well it is used in the majority of boilies whether home made or retail and it is used because it is cheap and acts as a brilliant binder helping to hold all the other ingredients together. It also takes on flavourings and colourings very well. Goodness wise it isn't high on the list but it does contain nutrients that fish need and adds to the value of the boilie

SOYA FLOUR
Soya flour is again a good binder and has a reasonable protein content which will add to the attraction of the bait. It also takes on flavours and colours very well.

MILK POWDER
Milk Powder is full of vitamins, minerals and proteins providing some great attractants to the fish. It also provides a smooth creamy finish to your boilie aiding in the rolling of it.

FISHMEAL
There are lots of different types of Fishmeal available some will have higher oil content, some will be more digestible however, the main reason for including Fishmeal in a boilie mix is that it has high amino acid and protein levels making the bait highly attractive to fish. One word of warning is you could have the opposite effect in winter if you use a high oil content Fishmeal as it will trap all the goodness, flavours and attractants inside the boilie. If you want to use Fishmeal in Winter I would advise using White Fishmeal as it has lower oil content and is an easy to digest Fishmeal.

EGG ALBUMEN
Egg Albumen is basically egg white powder and has two main reasons for being included in a boilie the first of which is to add a good protein level into a mix. The second reason is more to do with the construction of the boilie and helps to harden your boilie and also create a good shell on the outside of the boilie.

WHEY GEL
Whey Gel has one main purpose in boilie making and that is too eliminate the need for liquid eggs in a boilie, it can be boiled/steamed for less time than liquid eggs typically being around 60 seconds which means that more nutrients and flavourings will be kept in your boilies. If you want to remove liquid egg altogether from your recipe then include 5% Whey Gel and 5% Egg Albumin in your overall mix.

CORN GRIT
Corn Grit is an amazing ingredient to add into any of your bait mixes, it is a highly digestible corn meal that will give your bait bright dots of colour throughout your mix. The texture allows for the release of its own and other flavours more easily from your bait. It has a higher protein content than Fishmeal (and smells nicer). If you can't find Corn Grit from places such as a health food shop or local Asian grocers then search for a product called Supergold 60, although this will be more expensive.

WHEATGERM
Wheatgerm is included in baits at quite a low level as it is quite a buoyant, it also helps as a digestive aid so is a valuable ingredient to add into baits all year round but especially in winter baits. There are lots of Proteins and vitamins in Wheatgerm again adding to the attraction of the bait you are creating.

BLOOD POWDER
Blood Powder is something I don't use very often but when I do use it, it is to enhance the nutrient levels of a

bait as well as to help harden the bait. As a side effect you also get a nice dark Red colour to your bait.

BETAINE
Betaine is used in a lot of my boilies at an inclusion rate of around 1-2tsp per 16oz as it is quite a strong ingredient. I include it in my boilies as it helps digestion hugely, helps with the passing of nutrients through the fishes intestinal walls (a lot of people don't know that fish don't technically have a stomach and that most of the dietary and food requirements are absorbed through the walls of their intestines as it passes through them, this is the main reason anglers try to make their boilies as digestible as possible) and is also quite a sweet ingredient as it is a crystal that is extracted from Sugar Beet while making sugar.

BREWERS YEAST (Debittered)
Brewer's Yeast is a fantastic source of vitamins, minerals and will add huge attraction to your bait. I personally don't include huge quantities of this as it also acts as a binder and if included at too high a level will result in your bait mixture becoming stiff and difficult to roll or manipulate into different shapes.

KELP/SEAWEED POWDER
Seaweed Powder is another ingredient that is absolutely full of vitamins and minerals with the added benefit of naturally containing Betaine. You may not see this in my boilie mixes very often and this is for the simple reason it stinks when you boil/steam it lol.

Hopefully the explanation of some of the ingredients I use will help you see why they are an important part of my boilie constructions and also will give you some thought provoking ideas as too other ingredients you can include. The main thing to remember is that fish especially Carp will seek out food that is easy to digest for them and will also give them good nutrient, amino and vitamin consumption.

UNUSUAL BAIT INGREDIENTS

ACORN FLOUR

Okay so this one is one that I have introduced to bait making, its origins are actually from the health food market and is full of vitamins, minerals and omega 6 fatty acids all of which are not only healthy to humans but also to the majority of fish.

Firstly you will need to collect your acorns, now this is fairly easy to do but watch out for acorns that have a damaged shell, especially ones that have a dark hole or small circular scar about 4-5mm in diameter as these ones will have worms in them and are no good to use.

You will now need to dry out your acorns and the best way to do this is by removing the shell and leaving the soft fleshy nut in the centre. You can aid the drying process by using a dehydrator (follow your dehydrators instructions) or if you don't have a dehydrator allow them to get lots of air as you don't want them to go rotten.

Once the acorns are dry you will need to grind them I usually put them in a blender and add plenty of water. Don't worry about how much water you are using as you will need to rinse the acorns off several times anyway to remove the one harmful ingredient which is the tannins. Acorns contain a high amount of tannic acid which is very bitter tasting and can be harmful to kidneys or mineral absorption, it is however, water soluble so is easily removed.

To remove the tannin, place two sheets of cheese cloth over a large sieve and pour your acorn liquid from the blender into the cheese cloth allow the excess liquid to drain off then rinse with warm water for around two minutes again allowing the excess to drain off after the two minutes. Bring the edges of the cheese cloth together and squeeze out as much liquid as possible. Give the strained acorns a quick taste if it is still bitter repeat the rinsing process until the bitterness has gone.

You now need to dry out your acorn flour. Spread out the mixture on to a baking tray and either place in to a dehydrator if you have one or warm the oven up to 200 degrees Celsius then turn it off and put the baking tray in the oven to dry out as the oven cools down. Stir the mix occasionally to help speed up the process, if it starts to clump up then you are doing it right.

NETTLE POWDER

Another out there addition to my bait making armoury is nettle powder, this is made from what many of you would consider to be a weed, in actual fact it has many medicinal and nutritious qualities and has been used for centuries as food and medicine. The common stinging nettle is found in abundance all around the world.

Nettles are full of Iron, potassium, vitamin A, vitamin C and once they are made into powder they are around 40% protein. To get the best out of nettles it is best to pick them as younger plants that are up to 7 inches tall, remember to use gloves when picking them or you're going to get a lot of stings. Cut off the plant at ground level and then give them a rinse under cold water to remove any dirt or insects that may be attached to them. To dry them out take the leaves off the stem and place them all on large baking trays or dehydrator trays, you will need to place them in a low heat for 10-12 hours. The heat will not only dry out your nettle stems and leaves but it will also get rid of the sting.

Once your nettles have dried out simply grind them up in a coffee grinder or even with a good old pestle and mortar, place in to an air tight jar or container and the powder will keep its nutrients and proteins for at least 10 months.

Quantities vary depending on what bait you are making but if you are making boilies I use between 2 and 3 ounces in a 16 ounce mix. The best time to pick nettles is between February and March.

DANDELION SYRUP AND POWDER

Dandelions are another of those secret health foods that have been forgotten about in the modern world and are now classed as a weed and a pest. Dandelion leaves taste delicious in salads and are a good alternative to lettuce leaves or spinach. Dandelions are full of vitamins, minerals and nutrients and have been associated with helping diabetes, cirrhosis of the liver and has even been mentioned in treating cancer although there is no official statement on the last one.

Right, then so how can dandelions be used in bait making? Well there are two main parts that I use in my bait making and these are the yellow heads and the roots. Firstly I will tell you how to make a delicious tasting syrup that can be used to either dip baits in or used as a flavouring in your baits. So to start with you will need the following items in order to make your syrup...

65 Dandelion Yellow Flowers (careful they stain)
8oz Mineral Water
2oz Granulated Sugar
Juice from Quarter of a Lemon
Large Saucepan with Lid
Large Jug
Sieve\Strainer
Wooden Spoon
Jar with Air Tight Lid

The first thing you need to do is to remove any remaining stalk and\or leaves so that you only have the yellow flower left. Give them a quick rinse under water to remove any insects or pesticides that may have gotten on to the flower. Place all of the flowers into a large saucepan add the juice of the quarter lemon and then pour the water over the top.

Bring the contents of the saucepan to a boil, then reduce the heat to a simmer put the lid on and simmer for 60 minutes. You now need to remove from the heat and allow the mixture to sit for 14 hours. You now need to place the sieve over the large jug and drain the liquid into the jug leaving the dandelion heads in the sieve, discard the heads and return the liquid to the saucepan.

Add the 2oz of sugar to the dandelion liquid and again bring the mixture up to a boil stirring occasionally. Reduce the heat down to a simmer, put the lid on and simmer occasionally for about 90 minutes or until you have the thickness you want. Stir every now and then to stop it sticking to the bottom of the saucepan. Once it is cooked pour it in to the air tight jar, allow to cool, put the lid on and store in the fridge. If you want to use this as a flavouring you will need to use around 30ml per 500g of base mix for boilies and around 20ml for things such as jelly baits.

The next part of the dandelion to be used is the root and this is used to create dandelion powder. To make this you will need the following...

1lb Large Dandelion Roots
Large Sharp Knife
Large Bowl
Food Blender
Baking Trays
Oven

Take all of your dandelion roots and put them in a large bucket, fill with water and stir them around with your hands for a few minutes, the water should go very muddy so tip the water out and put fresh water in, repeat this process until the water is still clear after the 2 minutes. Don't worry if they aren't completely clean they will be washed again later on in the recipe.

Cut your roots in to large chunks around an inch in size then add a handful of roots to your blender and blend them on high power until they resemble coffee granules. Tip them out in to your bowl and repeat with the rest of the roots until you have ground them all.

Now spread your ground roots out onto your baking trays at around 2cm deep. Now you will need to set your oven on 250 degrees Celsius, put the trays in the oven, every now and then open the oven door and give them a stir this will give even cooking\roasting but will also allow any moisture to escape when you open the door. Cooking times will vary but will take roughly 2 hours, you are aiming at getting a dark coffee colour be careful not to burn them or they will be ruined. Remove from the oven and allow to cool, store in a large glass jar with an air tight lid. To make the roasted roots into a powder simply grind them down in a coffee grinder, I use around 1oz of dandelion powder per 16oz of base mix or 1 tsp in other baits.

You may have noticed I mentioned coffee a lot in this recipe and you can use these roasted dandelion roots to create a delicious tasting dandelion coffee. As before grind the roots in the coffee grinder but instead of using in a bait add 4 tablespoons (adjust to taste) of dandelion powder into a pint of water add this to either a saucepan or one of those kettles you can boil on a stove. Boil the dandelion coffee for around 15-20 minutes then strain through a coffee filter and serve as you would normally serve your coffee.

BASIC BOILIE RECIPES

Although I will be including a large selection of my boilie recipes in the following pages please do not feel you have to stick to these recipes. These recipes are here to help you get started with your bait making and although they have all caught fish don't forget to experiment with your own recipes too. Just remember the simple rule of fish welfare and don't include ingredients that are going to be unhealthy for the fish.

This first selection of recipes are my more basic recipes, this doesn't mean that they won't work but it does mean that after a while the fish may decide to they no longer want to eat them and you will need to change to a different bait.

MEAT BOILIES
6oz Semolina
4oz Soya Flour
2oz Wheat Flour
2oz Crushed Hemp Seed
2oz Ground Bait-Tech Marine Halibut Pellets
3 Large Eggs
1 Tin Bait-Tech Halibut Meat (liquidized or grated)
5ml Bait-Tech Chilli Oil
2tsp Brewer's Yeast
1 tsp Salt

pg. 48

CATFOOD BOILIE
6oz Semolina
4oz Soya Flour
2oz Cornmeal
2oz Rice Flour
1oz Ground Dry Catfood (the biscuit type)
1oz Crushed Hemp Seed
3 Large Eggs
30ml Liquidized Catfood (preferably a fish flavoured one)
2tsp Brewer's Yeast
2tsp Ground Rock Salt

ENERGY DRINK BOILIES
6oz Semolina
6oz Soya Flour
4oz Crushed Bait-Tech Marine Halibut Pellets
2 Large eggs
1 tsp Ground Rock Salt
50ml Energy Drink

HALIBUT & BETAINE BOILIES
6oz Semolina
4oz Soya Flour
3oz Ground Rice
2oz Cornmeal
1oz Fishmeal
3 Large Eggs
20ml Halibut Oil
2tbsp Betaine Powder
1tsp Brewer's Yeast
1tsp Sea Salt
25ml Black Food Colouring

COCONUT CREAM BOILIE
6oz Semolina
4oz Soya Flour
2oz Milk Powder
1oz Crushed Hemp
1oz Coconut Flour
2oz Horlicks
3 Large Eggs
3tsp Desiccated Coconut
50ml Condensed Milk
5ml Cod Liver Oil
2.5ml Vanilla Flavouring
2.5ml Almond Flavouring
5ml Bait-Tech CSL
1tsp Salt

SARDINE & TUNA BOILIES
6oz Semolina
4oz Soya Flour
3oz Fishmeal
1oz Cornmeal
1oz Chilli Powder
1oz Crushed Hemp Seed
3 Large Eggs
1 Tin Sardines in Tomato Sauce (Liquidized)
1 Tin Tuna in Brine (Liquidized)
3 drops Black Peppercorn Oil
1tsp Brewer's Yeast
1tsp Betaine
1tsp Ground Rock Salt
5ml Oyster Sauce

SPICY KRILL & SNAIL BOILIES
6oz Semolina
3oz Soya Flour
3oz Bait-Tech N-Hance Fishmeal
1oz Krill Powder
1oz Cornmeal
1oz Brewer's Yeast
1oz Chilli Powder
3 Large Eggs
25g Liquidized Water Snails
10ml Bait-Tech Omega3 Fish Oil
10ml Bait-Tech CSL Natural
2tsp Sugar
2tsp Betaine
1tsp Ground Rock Salt
25ml Black Food Colouring

RUM & COCONUT BOILIES
6oz Semolina
4oz Cornmeal
4oz Wheat Flour
1oz Desiccated Coconut
1oz SMA Gold Baby Milk Powder
3 Large Eggs
15ml Bait-Tech Sweet Coconut Flavour
10ml Asda Rum Flavouring
5ml Glycerine
1tsp Brewer's Yeast
1tsp Salt

MUSSEL BOILIES
6oz Semolina
4oz Soya Flour
4oz Fishmeal
1oz Seaweed Powder
1oz Krill Powder
2 Large Eggs
10ml Oyster Sauce
1tsp Sea Salt
150g fresh liquidized mussels including finely crushed shells

SQUID & SCOPEX BOILIES
6oz Semolina
4oz Soya Flour
1oz Ground Rice
5oz Squid Meal
3 Large Eggs
10ml Scopex (20ml if it's Dynamite squeezy bottle)
10ml Cod Liver Oil
1tsp Brewer's Yeast
1tsp Salt
25ml Red Food Colouring

OYSTER BOILIES
6oz Semolina
2oz Soya Flour
4oz Tiger nut Flour
3oz Fine Oyster Shell
1oz Krill Powder
3 Large Eggs
20ml Oyster Sauce
5ml Bait-Tech CSL
1tsp Sea Salt

PEANUT SPICE BOILIES
6oz Semolina
2oz Soya Flour
2oz Cornmeal
2oz Tiger nut Flour
2oz Peanut Flour (can make yourself by grinding peanuts in a coffee grinder)
2oz Coarsely Crushed Peanuts (soaked for 24hrs and Boiled for 30mins)
3 Large Eggs
1tsp Salt
1tbsp Chilli Powder
10ml Corn Steep Liquor
10ml Peanut Flavouring (5ml Peanut Oil)

WILDBERRY BOILIES
6oz Semolina
4oz Soya Flour
2oz Tiger nut Flour
2oz Crushed Hemp
2oz SMA Gold Baby Milk Powder
3 Large Eggs
10 Blackberries Liquidized
10 Raspberries Liquidized
10 Blueberries Liquidized
5 Strawberries Liquidized
20ml Glycerine

MACKEREL BOILIES
5oz Semolina
4oz Soya Flour
6oz Fishmeal
1oz Krill Powder
3 Large Eggs
15ml Oyster Sauce
15ml Bait-Tech Omega 3 Fish Oil
10ml Mackerel Flavouring (or fresh liquidized Mackerel)

FRUITY TIGER BOILIES
6oz Semolina
4oz Soya Flour
4oz Tiger nut Flour
1oz Ground Tiger Nuts
1oz 5Pints Milk Powder
3 Large Eggs
5ml Bait-Tech CSL
2.5ml Glycerine
5ml Tutti Frutti Flavouring
2tsp Sea Salt
1tsp Brewer's Yeast

DANDELION & ELDERBERRY BOILIE
6oz Semolina
4oz Soya Flour
3oz Wheat Flour
2oz Cornmeal
1oz Brewer's Yeast
3 Large Eggs
1oz Dandelion Leaves (weigh then liquidize)
1oz Elderberries (weigh then liquidize)
25ml Glycerine
4drops Black Peppercorn Oil

1tbsp Clear Honey
3tsp Chilli Powder
1tsp Ground Rock Salt
1tsp Chives, Freshly Chopped
1tsp Spirulina

SWEETCORN BOILIE
6oz Semolina
4oz Soya Flour
4oz Cornmeal
1oz SMA Gold (baby milk powder)
1oz Crushed Hemp
3 Large Eggs
20ml Bait-Tech CSL
5ml Oyster Sauce
1 Small Tin Sweetcorn Liquidized
10ml Glycerine
1tsp Chilli Powder
25ml Yellow Food Colouring

THE ORANGE BOILIE
6oz Semolina
4oz Soya Flour
3oz Wheat Flour
2oz 5Pints Milk Powder
1oz Krill Powder
2 Large Eggs
50ml Concentrated Orange Juice (with or without bits)
3 drops Orange Essential Oil
2tbsp Artificial Sweetener
25ml Orange Food Colouring

SPICY PRAWN BOILIES
6oz Semolina
4oz Soya Flour
2oz Fishmeal
2oz Wheat Flour
1oz Krill Powder
1oz Chilli Powder
3 Large Eggs
25 Cooked Peeled Prawns Liquidized
10ml Glycerine
2 tbsp Mayonnaise
1 tbsp Tomato Ketchup
1 tsp Oyster Sauce
1 tsp Tabasco Sauce
1 tsp Brewer's Yeast
1 tsp Ground Rock Salt
15ml Pink Food Colouring

TANGERINE & PEACH BOILIES
4oz Semolina
5oz Soya Flour
1oz Tiger Nut Flour
1oz Crushed Hemp
1oz ground Rice
2oz 5pints Milk Powder
2oz Cornmeal
3 Large Eggs
1tsp Chilli Powder
5ml Bait-Tech CSL
5ml Glycerine
2.5ml Tangerine Flavouring
2.5ml Peach Flavouring
25ml Orange Food Colouring
5ml White Food Dye (available from Mainline)

LIVER BOILIES
5oz Semolina
4oz Soya Flour
4oz Fishmeal
3oz Blood Powder
2 Large Eggs with crushed shells
150g Liver (liquidized)
5ml Bait-Tech CSL
5ml Oyster Sauce
1 tsp Garlic Powder (not salt)
3 tsp Clear Honey
1 tsp Brewer's Yeast

BETTER BOILIE RECIPES

These recipes are slightly more advanced than the basic recipes and will last for a longer period of time before the fish go searching for another source of food.

ANTS SPECIALS BOILIE
4oz Semolina
4oz Soya Flour
2oz Cornmeal
2oz SMA Gold
2oz Krill Powder
1oz Crushed Hemp
1oz Fishmeal
4 Large Eggs
1tsp Powdered Fenugreek
2 Cubes Frozen Bloodworm (defrosted)
1tbsp Crushed Water Snails
1tsp Ground Rock Salt

1tsp Brewer's Yeast
1tbsp Tomato Ketchup
1tsp Garlic Granules
4 Drops Black Peppercorn Oil
10ml Oyster Sauce
15ml Bait-Tech Omega 3 Oil
25ml Black Food Colouring

TOFFEE & BANANA BOILIES
6oz Semolina
3oz Soya Flour
2oz SMA Gold Milk Powder
2oz Ground Rice
1oz Banana Milkshake
1oz Tiger Nut Flour
1oz Crushed Hemp Seed
2 Large Eggs
Half a Ripe Banana (Liquidized)
50ml Condensed Milk
10ml Glycerine
5ml Toffee Flavouring
5ml Bait-Tech CSL Natural
2tsp Brewer's Yeast
1tsp Ground Rock Salt

SQUID BOILIES
6oz Semolina
3oz Soya Flour
3oz Squid Meal
1oz Fish Meal
1oz SMA Gold
1oz Robin Red
1oz Crushed Hemp Seed
3 Large Eggs

Half a Liquidized Banana
15ml Bait-Tech CSL
3ml Squid Flavouring
2ml Scopex Flavouring
3tsp Banana Milkshake Powder
2tsp Brewer's Yeast
2tsp Ground Rock Salt
1tsp Black Mainline Tru-Colour Powder

BFG BOILIE
6oz Semolina
4oz Soya Flour
2oz Ground Rice
1oz Cornmeal
1oz Krill Powder
1oz Pre-Digested Fishmeal
1oz Brewer's Yeast
3 Large Eggs
20ml Beetroot Juice
2tsp Fructose
1tbsp Golden Syrup
3tbsp Marmite
1tsp Ground Rock Salt
3tsp Ground Dried Mealworm
5ml Bait-Tech CSL
5ml Oyster Sauce
5ml Liquid Molasses

WORM COCKTAIL BOILIES
7oz Fishmeal
2.5oz Semolina
1.5oz Rice Flour
1.5oz Soya Flour
1.5oz Milk Powder

1oz Wheatgerm
0.5oz Betaine
0.5oz Horlicks
3 Large Eggs
10ml CC Moore Worm Extract
10ml Sticky Baits Bloodworm Extract
5 Liquidized Lobworms
2 Cubes of Frozen Bloodworm (defrosted)
1tsp Ground Rock salt

LONG TERM BOILIE RECIPES

These are my more advanced recipes and will last you for season after season. Please take a moment to look at the layout of the recipes, the ingredients that are listed and you will start to see how you can start making your own long term recipes.

TOFFEE CREAM BOILIES
4oz Haith's CLO
4oz Lamlac
1.6oz Semolina
1.6oz Vanilla Meal
1.6oz Ground Rice
0.8oz Whole Egg Powder
0.8oz Wheatgerm
0.8oz Horlicks
0.4oz Egg Albumin
0.4oz Ground Hemp Seed
2 Large Eggs
50ml Double Cream
10ml Bait-Tech CSL

5ml Toffee Flavouring
2tsp Brewer's Yeast
2tsp Fructose

BAILEYS BOILIES
4oz Haith's CLO
4oz Vanilla Meal
1.6oz Semolina
1.6oz Tiger Nut Flour
1.6oz Whey Powder
0.8oz Whole Egg Powder
0.8oz Wheatgerm
0.8oz CC Moore SuperGold 60
0.4oz Egg Albumin
0.4oz Ground Hemp Seed
2 Large Eggs
40ml Baileys Irish Cream (Original)
10ml Hemp Oil
1.5ml Almond Essence

ANISEED BOILIES
4oz Haith's CLO
4oz Haith's SuperRed
1.6oz Semolina
1.6oz Lamlac
1.6oz Tiger Nut Flour
0.8oz Krill Meal
0.8oz Blood Powder
0.8oz Whole Egg Powder
0.4oz Seaweed Meal
0.4oz Egg Albumin
3 Large Eggs
10ml Bait-Tech CSL
5ml Aniseed Flavouring

2tsp Ground Fennel
2tsp Fructose
1tsp Hemp Protein Powder
1tsp Powdered Red Food Colouring

WHITE CHOCOLATE BOILIES
4oz Haith's CLO
4oz White Semolina
1.6oz Horlicks
1.6oz Vanilla Meal
1.6oz Lamlac
0.8oz Whole Egg Powder
0.8oz Wheatgerm
0.8oz Whey Powder
0.4oz Egg Albumin
0.4oz Ground White Rice
3 Large Eggs
70g Finely Grated White Chocolate
20ml White Chocolate Syrup
2tsp Titanium Dioxide

SANAGI BOILIES
4oz Marukyu SFA441 Sanagi Powder
4oz Haith's CLO
1.6oz Semolina
1.6oz LT94 Fishmeal
1.6oz Haith's Robin Gold
0.8oz Whole Egg Powder
0.8oz Seaweed Meal
0.8oz Haith's SuperRed
0.4oz Blood Powder
0.4oz Wheatgerm
3 Large Eggs
20ml Sanagi Liquid Extract (Marukyu SFA430)

10ml Hemp Oil
2tsp Ground Sea Salt
2tsp Brewer's Yeast

LIQUORICE BOILIES
4oz Haith's CLO
4oz Haith's Insectivorous
1.6oz Semolina
1.6oz Lamlac
1.6oz Finely Ground Molasses Meal
0.8oz Ground Rice
0.8oz Whole Egg Powder
0.8oz Wheatgerm
0.4oz Blood Powder
0.4oz Egg Albumin
2 Large Eggs
30ml Liquid Molasses
20ml Holland & Barrett Liquid Amino
2tsp Ground Fennel
1tsp Powdered Anise
1tsp Powdered Liquorice Root
2tsp Black Powdered Food Colouring

SCOPEX, SQUID & LIVER BOILIES
4oz Squid Meal
4oz Haith's CLO
1.6oz Semolina
1.6oz LT94 Fishmeal
1.6oz Haith's Robin Red
0.8oz Wheatgerm
0.8oz Blood Powder
0.8oz CC Moore SuperGold 60
0.4oz Seaweed Meal
0.4oz Ground Halibut Pellet Powder

1 Large Egg
100ml Mineral Water
15g Liver Powder
10ml Squid Ink
5ml Scopex Flavouring
2tsp Brewer's Yeast
1tsp Sea Salt

PEANUT BUTTER BOILIE
4oz Haith's CLO
4oz Vanilla Meal
1.6oz Ground Oats
1.6oz Semolina
1.6oz Peanut Flour
0.8oz Wheatgerm
0.8oz Casein Powder
0.8oz Whey Powder
0.4oz Lamlac
0.4oz Tiger Nut Flour
2 Large Eggs
50ml Almond Milk
70g Smooth Peanut Butter
3tsp Fructose
2tsp Brewer's Yeast
1tsp Hemp Protein

HOTDOG BOILIES
4oz Haith's CLO
4oz Cornmeal
1.6oz Semolina
1.6oz Potato Flour
1.6oz Lamlac
0.8oz Wheatgerm
0.8oz Whey Powder

0.8oz Rice Flour
0.4oz Seaweed Meal
0.4oz Blood Powder
3 Large Eggs
20ml Holland & Barrett Liquid Amino
5ml Spicy Meat Flavouring
2tsp Mustard Powder
2tsp Coriander
1tsp Paprika
1tsp Garlic Powder
1tsp Black Pepper

NATURALS BOILIE
4oz Haith's CLO
4oz Haith's Insectivorous
1.6oz Semolina
1.6oz Tiger Nut Flour
1.6oz LT94 Fishmeal
0.8oz Ground Hemp Seed
0.8oz Ground Halibut Pellet Powder
0.8oz Krill Meal
0.4oz Seaweed Meal
0.4oz Wheatgerm
2 Large Eggs
2 Cubes Frozen Bloodworm (defrosted)
1tbsp Crushed Water Snails
10ml Lobworm Extract
10ml Bloodworm Extract
10ml Liquid Krill
10ml Oyster Sauce
10ml Bait-Tech Omega 3 Oil
2tsp Sea Salt

RED SEA BOILIES

4oz LT94 Fishmeal
4oz Haith's CLO
1.6oz Semolina
1.6oz Haith's Robin Red
1.6oz Krill Meal
0.8oz Seaweed Meal
0.8oz Fine Oyster Shell
0.8oz Wheatgerm
0.4oz Blood Powder
0.4oz Ground Hemp Seed
3 Large Eggs
20ml Oyster Sauce
10ml Cod Liver Oil
2tsp Paprika
2tsp Brewer's Yeast
1tsp Ground Rock Salt
2tsp Red Powdered Food Colouring

HALIBUT & HEMP BOILIES
4oz Haith's CLO
4oz Ground Halibut Pellet Powder
1.6oz Semolina
1.6oz Ground Hemp Seeds
1.6oz Lamlac
0.8oz Blood Powder
0.8oz Whole Egg Powder
0.8oz Wheatgerm
0.4oz Haith's SuperRed
0.4oz Seaweed Meal
3 Large Eggs
15ml Hemp Oil
5ml Omega 3 Fish Oils
2tsp Hemp Protein Powder
1tsp Garlic Powder

1tsp Brewer's Yeast
1tsp Ground Rock Salt

TUNA BOILIES
4oz LT94 Fishmeal
4oz Haith's CLO
1.6oz Semolina
1.6oz Lamlac
1.6oz CC Moore SuperGold 60
0.8oz Wheatgerm
0.8oz Ground Halibut Pellet Powder
0.8oz Krill Meal
0.4oz Blood Powder
0.4oz Seaweed Meal
1 Large Egg
100ml Mineral Water
10ml Squid Ink
10ml Bait-Tech Omega 3 Fish Oil
1 x Tin of Tuna in Brine
2tsp Chilli Powder
2tsp Paprika
2tsp Brewer's Yeast
2tsp Ground Rock Salt

BLOODWORM BOILIES
4oz LT94 Fishmeal
4oz Haith's CLO
1.6oz Semolina
1.6oz Krill Meal
1.6oz Pre-Digested Fishmeal
0.8oz Whole Egg Powder
0.8oz Fine Oyster Shell
0.8oz Blood Powder
0.4oz Ground Hemp Seed

0.4oz Seaweed Meal
3 Large Eggs
20ml Sticky Baits Bloodworm Extract
10ml Oyster Sauce
10ml Bait-Tech Omega 3 Fish Oil
1tsp Brewer's Yeast
1tsp Ground Rock Salt
1tsp Hemp Protein Powder

SPICY MANGO & COCONUT BOILIES
4oz Haith's CLO
4oz CC Moore Vanilla Meal
1.6oz Semolina
1.6oz Lamlac
1.6oz Coconut Flour
1.6oz Wheatgerm
0.8oz Rice Flour
0.4oz Brewer's Yeast
0.4oz Ground Hemp Seed
1 Large Eggs
70ml Mango Juice
30ml Coconut Milk
2tsp Hot Chilli Powder
2tsp Paprika
1tsp Fructose
2tsp Orange Food Colouring Powder

SPICY TOMATO BOILIES
4oz Haith's CLO
4oz Haith's SuperRed
1.6oz Semolina
1.6oz Lamlac
1.6oz Ground Rice
0.8oz Whole Egg Powder

0.8oz Wheatgerm
0.8oz CC Moore SuperGold 60
0.4oz Brewer's Yeast
0.4oz Chilli Powder
1 Large Egg including crushed shell
100ml Water
60ml Tomato Purée
1tsp Betaine
1tsp Sea Salt
1tsp Coarse Ground Black Pepper
1tsp Paprika
1 Multivitamin Tablet Crushed
1tsp Red Powder Food Colouring

CREAMY CHOCOLATE AND CHILLI BOILIES
4oz Haith's CLO
4oz Semolina
1.6oz Hot Chocolate Powder
1.6oz Tiger Nut Flour
1.6oz Vanilla Meal
0.8oz Horlicks
0.8oz Whey Powder
0.8oz Wheatgerm
0.4oz Ground Hemp Seed
0.4oz Chilli Powder
2 Large Eggs
30ml Double Cream
20ml Holland & Barrett Liquid Amino
15g Finely Grated Cooking Chocolate
2tsp Brewer's Yeast
1tsp Fructose

HOME MADE POP UPS

I have been asked on several occasions if I will write a chapter on different ways to make pop up boilies as there isn't many out there or if there were they have been lost in the long distant past.

Okay where to begin, well in the eighties pop ups used to be made by combining 2oz Sodium Casienate, 2oz Calcium Casienate and 1oz Soya Isolate with 1 Large egg and 10ml of flavouring creating a highly attractive protein rich pop up.

However, milk proteins became very expensive to buy and to this day are still on the expensive side so people started experimenting with other ideas the first of which was to wrap a thin layer of your normal boilie mix around a polyball in order to make it float.

Wrapping a layer of boilie paste around a corkball is also a popular method still used to this day. I personally don't like using corkballs as if it does happen to come off the hair it could get stuck in the fish's intestines and cause problems although this is a very rare occurrence.

Another successful although less reliable method is to microwave your normal boilie mix instead of boiling it. If you use the following method you will get pop ups but results will vary as to how long they will stay buoyant. The base mix is prepared in the normal way and rolled into balls as normal, however instead of boiling the baits microwave them instead.

Typically microwave 30 x 14mm boilies at a time. They are placed on a microwave type dish or plate and microwaved on full power for 2 minutes. At the end of the 2 minutes examine the baits for any sign of burning, if there is none continue to microwave the baits on full power for a further 15 seconds.

Again examine the baits for any signs of burning, continue to microwave the baits in 15 second bursts examining the baits each time for any slight burn marks. As soon as you detect any signs of burning, stop the process and you now have your pop-ups.

The longer you can microwave the better they pop up but don't burn them. If you want different sizes than 14mm you will need to experiment with timings yourself as I've only ever done 14mm. One last thing on microwaving boilies the more milk based product such as casein you include the more buoyant they will be.

Another very successful way to make your own pop ups is to buy one of the many ready made pop up mixes available from companies such as Mainline, Richworth, CC Moore and many more. All are excellent quality and will produce you a good pop up, one good tip for these is for every 4tbsp of pop up mix you use add 1tbsp of egg albumin, it helps them roll better.

Right we are now onto my favourite way of making pop ups... Cork Dust. To make cork dust pop ups you will need to go by volume rather than weight. I always use the following measurements as a starting guideline for a one egg mix although you may end up needing more or less mix depending on the size of your egg.

Okay so you will need;
One level coffee mug of your chosen base mix
One level coffee mug of cork dust
Half a coffee mug of egg albumin (very important as it helps bind)
One large egg
5g potassium sorbate diluted in 10ml warm water (makes them shelf life)
5ml concentrated bait flavouring
1tsp coloured powdered food dye

If you use any other additives such as brewer's yeast, salt, fructose, etc add it to the base mix before you measure out the coffee cup of it.

Now tip all of the ingredients from the coffee cup (use the same coffee cup to measure out each ingredient) into a large air tight sealable bag, seal it and give it a good shake to mix everything together.

Next mix your egg, food dye, potassium sorbate and flavouring in a medium mixing bowl and slowly start adding your mixed dry ingredients until you have a dough that just doesn't stick to your hands.

Use your hands or a sausage gun to create your pop up sausages then either break bits of and roll into balls or use a rolling table to create more uniform sized pop ups.

Finally boil around 30 pop ups at a time for 120 seconds and allow them to air dry for 48 hours before putting them into air tight containers/jars. Once every three weeks give them a couple of sprays with a fine mist sprayer of the flavouring you used just to keep them flavoured up.

On a last note it is worth remembering that pop ups become less effective in deeper water. At around 10ft the pop up will start to become less buoyant and will become less effective as it gets deeper basically becoming a none buoyant bottom bait at around 30ft.

HOME MADE PASTES

So in this chapter I will show you how to make paste there are a few types I use and I will tell you how to make them. Right let's get started...

The first paste that I use is based on a very simple recipe which is as follows;
2oz Wheat Flour
4oz Rice Flour
160ml Water
10ml Colouring
5ml Flavouring
1tsp Salt

The first thing to do is stir your flavouring and colouring in to the water and set it to one side. Now put your wheat flour and salt into a mixing bowl and slowly pour in your water a bit at a time stirring constantly with a fork until your mix ends up as a stiff lump around your fork. (You might need slightly more/less water to the recipe to achieve this)

pg. 73

Put a little bit of flour on a clean surface and move your mix to this surface. Now using your hands keep folding the mix in on itself for about 5 minutes until you have a paste that you can move around without breaking it. Keep your paste in a plastic air tight bag and you should be okay for about a week. If you want a paste that lasts for a very long time in the water simply use your boilie mix when it is at the playdough stage.

The other paste I like to use is a bit of an oldie but it definitely gets some nice results and that is bread paste. To make bread paste is extremely easy, take a slice of bread and cut the crusts off (I use Kingsmill medium bread). Now take some oil I like to use the Bait-Tech X-Cite range of oils but you can use any type of oil and spread it on the bread like you would butter, leave it to stand for 3 minutes.

After 3 minutes, you need to keep squashing the bread in your hands until you have a paste, leaves a nice scent trail in the water and catches loads of fish. Don't overdo it with the oil or you will get a lot of wastage, but experiment and see what works best for you.

Finally I like to use pellet pastes, this type of paste has a much higher nutritional value to fish and can be used all year round.

So what do you need to make this paste? Well you need pellets, you also need to be careful which pellets you choose to use as for example in Winter High Oil Pellets are not as suitable as they will not breakdown as quickly and are not as easily digested. To prepare your pellets place your required amount in a saucepan as a guideline I usually use around 200g of pellets, at this stage I will point out that you can control the breakdown times of your pellets from minutes to hours depending on the type of pellet you use. Layers Pellets will breakdown very quickly whereas High Oil Halibut Pellets will have a lot longer breakdown time, one of the ways to alter your pastes breakdown time is to mix and match your pellets so you could have 50g Halibut Pellets and 150g Layers Pellets for a firm but quick dissolving paste.

If I am using any additives I will add them at this stage, here is my two favourite recipes to get you started but the ingredients you can use is only limited by your imagination...

Warm Weather Paste
200g Bait-Tech Marine Halibut Pellets
10ml Bait-Tech Omega3 Fish Oil
5ml Bait-Tech CSL
5g Betaine
5g Ground Meal Worm
5g Ground Rock Salt
5g Brewer's Yeast

Cold Weather Paste
100g Bait-Tech Fishmeal Carp Feed Pellets
100g Layers Pellets
10ml Salmon Oil
5ml Bait-Tech CSL
5ml Pineapple Flavouring
5g Betaine
5g Fructose
5g Chilli Powder

5g Brewer's Yeast

To make your paste put your pellets into the saucepan then add all your other ingredients and give them a good stir round to try and get all of the pellets coated. Allow them to rest for 30 minutes before pouring warm water over them until they are just covered.

Now for the hard part, it is difficult to give you an exact time to soak the pellets for but you should end up with all of the pellets soft and all the liquid soaked in.

The last stage is to mash them all into your paste and you are ready to use them.

HOME MADE JELLY BAITS

One of my favourite types of fishing is stalking in the margins and an ideal bait for margin work is another great classic which is the Jelly Bait, this is a fantastic bait that releases lots of flavours but due to its delicate nature won't survive a long cast. To make your jelly baits you will need the following ingredients

500ml Water
100g Gelatine
400g Base Mix/Groundbait of your choice
100g Sugar
2tsp Sweetener
5ml Flavouring
10g Potassium Sorbate

In a large saucepan, add the Water and the Gelatine and dissolve the Gelatine by bringing it all to a gentle simmer.

Once the Gelatine has dissolved and the mixture is hot add the Potassium Sorbate give a quick stir and then add the sugar stirring constantly until the sugar has dissolved. Pour all of your chosen base mix in and stir until well mixed.

Now pour the mix into a shallow tin that will allow it to set and be about 5cm thick. Place in fridge for at least an hour until set. Now you can take it out and cut it into cubes that are 3cm x 3cm. Perfect for margin work. The other option is to use round meat punches to cut out round baits.

These Jelly Baits are extremely soft so in order to stop the hair cutting through the bait you need to put a piece of plastic tubing on the hair and a small meat stop at the end. I can't remember where I got this recipe from but have used it for years with some great results.

Jelly Pellets are a fantastic bait and have the advantage over normal expander pellets that they are firmer and will stay on the hook a lot better. The first thing I would like to point out is that there are several different quality expander pellets out there, you can use this guide using any quality pellet, however I would recommend using Bait-Tech's Xpand Hook & Feed Pellets As these are good quality pellets and are not expensive at around £3 for a bag. You will need the following in order to make your Jelly Pellets... Gelatine, pellet pump, expander pellets, flavouring and water to make the Jelly.

As Gelatine is fairly cheap I tend to make more than I need to make sure there is enough to soak the pellets but if you want to work it out exactly you will need enough mixture to fill half of your chosen pellet pump.

The first step of making your Jelly Pellets is to fill your pellet pump a quarter full with dry expander pellets and set to one side. Now you will need to make up your Jelly as per the instructions on the Gelatine packet (try to make up about a pint), once all of the Gelatine crystals have dissolved add 5ml of your chosen flavouring and stir it in. If you wish to you can add 20ml of food colouring at this point as well.

Next you need to pour your jelly mix into the pellet pump over the dry expanders, fill it to half way full, put the lid on and then pump until it is really hard to push the pump down. Leave it like this for 10 seconds, then pull the valve that releases the pressure if you have done it right most if not all of the pellets will sink to the bottom of the liquid.

Lastly you will need to put all of the pellets onto a flat plate or baking tray and leave in the fridge overnight to set. You should find when you take them out the next day you have some nice firm Jelly Pellets, as with normal expanders hook them through the side and roll them onto your hook.

As a quick method, if you prefer to you can use a normal jelly mix such as Strawberry, Orange or any other flavours that these mixes come in, then you just make the mix and repeat the process from pouring over the pellets and pumping them.

The last jelly bait that I would like to show you how to make is Jelly Bread, this is really simple and although it can look a bit cumbersome fish absolutely love it.

In order to make Jelly Bread you will need the following ingredients. (You can add more if you wish)

Flavoured Jelly Mix
Loaf of Bread
2 Large Eggs

So on to the fun part, break the loaf of bread up into small pieces and put into a large bowl. It's with crustless but you get a bait not quite as firm.

Make up your jelly mix as instructed on the jelly packet and pour it over the bread (I've only tried Strawberry so far). Leave the bread to soak up all the Jelly.

Whisk the 2 eggs, pour into the mixture, then mix by hand until it's doughy.

Roll the mix in to boilie sized balls and place on a baking tray (or spread out on the tray). Bake in the oven at 190c for about 25-30 minutes. If you have spread the mix out on the tray you can cut it into cubes or use a meat punch to create pellet type baits.

HOME MADE GROUNDBAITS

One of the other things I like to make is groundbait, a groundbait is designed to be an attractor to your swim and can be complimented with a few freebies that match your hook bait.

The Basic Recipe
12oz Brown Bread Crumb
16oz White Bread Crumb
8oz Ground Hemp Seed
8oz Ground Halibut Pellet

Now to this you can add all sorts of little goodies but I wouldn't add any more than 3 out of these extra additives (or your own concoction).

Angel Delight 1 packet per quantity of above mix.
Custard Powder 1 packet per quantity of above mix.
Liquidized Boilies 30 per quantity of above mix
Ground Layer Pellets 8oz per quantity of above mix
Ground Trout Pellets 8oz per quantity of above mix
Fishmeal 8oz per quantity of above mix
Ground Birdseed 8oz per quantity of above mix
Krill Meal 4oz per quantity of above mix

I thought I would go into a bit more detail about groundbaits that I make and use for different species. A groundbait is designed to be an attractor to your swim and can be complimented with a few freebies that match your hook bait.

To prepare your groundbait firstly put all your dry ingredients into a big airtight bag, blow air into it and twist the top to seal. Give the bag a good shake to mix all the powders together, put your dry "groundbait" into whatever you are going to mix your groundbait in and add your liquid flavourings and give it a good mix round.

Finally you need to add small amounts of lake water until you have a mix that will stay together when squeezed once but will break up easily when rubbed between your hands. If you want to make a really fine groundbait riddle your groundbait through a sieve into another container now leave to settle for 30 minutes and then add around half a small sweetcorn tin of whichever hook bait you are using, then add more lake water if needed.

When you are making your groundbait balls make sure they are quite firm as you don't want them to break up as soon as they hit the water.

ROACH GROUNDBAIT
12oz Brown Breadcrumb
12oz White Breadcrumb
4oz Cornmeal
3oz Milk Powder
1oz Crushed Hemp
15ml Almond Food Flavouring
30ml Condensed Milk
2tsp Coriander Powder

BREAM GROUNDBAIT
12oz Brown Breadcrumb
12oz White Breadcrumb
4oz Cornmeal
3oz Molasses Meal
1oz Betaine
50ml Condensed Milk

TENCH GROUNDBAIT
12oz Red Breadcrumb
12oz White Breadcrumb
4oz Cornmeal
2oz Molasses Meal
1oz Blood Powder
1oz Mole Hill Soil (yes I mean exactly what it says lol)
1tsp Chilli Powder
Quarter of a small tin of Sweetcorn
Quarter of a tin of Castors

CARP GROUNDBAIT
12oz Brown Breadcrumb
12oz White Breadcrumb
4oz Cornmeal
2oz Fishmeal
1oz Ground Halibut Pellet
1oz Ground Hemp Seed
2tsp chilli powder
2tsp Cayan Pepper
Quarter of a small sweetcorn tin of 3mm Pellets
50ml Bait-Tech Omega 3 Fish Oil

BARBEL GROUNDBAIT
16oz Brown Breadcrumb
14oz Fishmeal

2oz Ground Hemp Seed
50ml Bait-Tech Omega 3 Fish Oil
A Handful of 6mm Halibut Pellets

COCONUT CREAM GROUNDBAIT
14oz Brown Breadcrumb
14oz White Breadcrumb
2oz Milk Powder
2oz Coconut Flour
2tbsp Desiccated Coconut
50ml Condensed Milk
5ml Vanilla Essence
5ml Almond Essence
Lake Water

You can make up the dry ingredients in your groundbait months in advance but it needs to be stored in the freezer, you can take out a couple of hours before your ready to use in order to defrost.

SPOD MIX & STICK MIX

Another great way of getting a small bunch of attractors around your hook bait is to use a small mixture of powders, particles, etc packed tightly into a 2-3inch PVA mesh bag. This is called a stick mix, my favourite stick mix recipe is as follows and has caught me quite a few fish.

5oz Bait-Tech Special "G" Groundbait
5oz Bait-Tech Halibut Marine Method Mix
1oz Krill Powder
1 tin of Anchovies in oil drained and liquidized
Bait-Tech X-Cite Omega 3 Fish Oil

Add all your ingredients except the oil into a large mixing bowl and give a thorough stir around to mix it all together. Now you want to squirt in the X-Cite Oil mixing it in until you have a texture that will form a good solid shape when squeezed in your hand but that will break down nicely when you rub your hands together. If you want to make your mix really smooth, riddle and then rub it through a fine mesh sieve once you have made it up. This mix should last for around a week and is PVA friendly.

I have been asked by a lot of people if I have any good Spod mixes, well I do have a couple that I use but I now use a Spomb instead of a Spod. So the first Spod mix is a hemp based Spod mix and will attract lots of different species but is also an easy to make mix which has given me some great results.

To start off you will need 900g prepared hemp seed (to prepare your hemp seed soak it in water for 24 hours or until it splits open), to this add a 460g tin of sweetcorn including the liquid, 1kg Bait-Tech Special 'G' Gold, 500g Bait-Tech Marine Halibut Method Mix, 1kg Bait-Tech 3mm Marine Halibut Pellets, 3tbsp Chilli Powder, small carton of Greek yoghurt and 3tbsp Salt. This mix is a slightly expensive one and is best used if you are after a specific fish or larger fish.

My next mix is a lot cheaper and will still successfully bring you fish into your swim. Start by soaking 3kg of Vitalin in warm water for 24 hours, after the first hour once the water has cooled down add 1kg Pigeon Conditioner into the Vitalin along with a tin of 460g tin of sweetcorn, 500g Groundbait and 500g Prepared Hemp Seed

PARTICLE PREPERATION

Something that I haven't covered lots is the preparation of particles. This is actually a very important thing to do as unprepared particles can be deadly to fish not only because some of them are toxic when unprepared but also the fact that the majority of them will swell to almost double their size which could burst a fishes stomach if too many are swallowed.

Okay the main particles that people want to prepare can be prepared in the same way and they are peanuts, French maize, maple peas, tiger nuts, whole brazil nuts, bird seed, sweetcorn and Sweet Lupins.

The first thing to do is place your chosen particles (you can mix and match) into a large saucepan, if you want to flavour them with either a liquid or powder additive add them now along with your colouring if you would like to colour them.

Next you want to cover the particles in water and make sure that you cover them by a good two inches to allow for the particles to expand. Give them a good stir, put the lid of the saucepan on and leave them to soak for twenty four hours.

Once they have soaked for twenty four hours, if you look at them you will notice that they have expanded some of them considerably depending on the particles you have used.

Bring the particles to a boil and allow them to boil for thirty minutes to soften them up. Once you have boiled them for the thirty minutes drain them off and set them to one side and allow them to cool.

Your particles are now ready to use, although you can if you want to place your particles into a coffee grinder and grind them down to a powder for adding into a bait mix or groundbait.

You can also use the exact method above for the following particles however you only need to soak for twelve hours and boil for thirty minutes so the particles are… soya beans, pinto beans, buck wheat, graded wheat, red kidney beans and broken brazil nuts. You can add these particles to any of the previously mentioned particles but always soak for twenty four hours if mixing the two different styles.

HOME MADE LUNCHEON MEAT

Luncheon Meat is a very popular bait and catches hundreds if not thousands of fish every week. With luncheon meat being so popular it is always worth trying something a bit different with it and there are lots of things you can do to boost the attraction of your luncheon meat.

The first is a very simple and easy trick, empty a can of regular Pepsi into a saucepan and bring it to a boil. Now while you are waiting for it to boil cut a tin of luncheon meat up into 8mm-12mm size cubes (you can do whatever size you like but these are what I use). Once the Pepsi is boiling tip the luncheon meat cubes in and reduce the heat, simmer for 5 minutes stirring roughly every 30 seconds, after 5 minutes remove the saucepan from the heat and drain the excess Pepsi off. Allow the luncheon meat to cool and you are ready to use.

The next method is another simple way to enhance your bait. Cut your luncheon meat up into cubes and place in a large mixing bowl, tip some icing sugar over it and gently mix it in with your hands being careful not to break the cubes of meat, shake off the excess icing sugar and discard it. Set the bowl with the luncheon meat to one side for about 5 minutes until you see oils coming out of the luncheon meat. Now take a dry powder such as groundbait or a spice and tip a good amount over the meat, again work it in with your hands. Pour the meat into a sieve and shake off the excess powders, you are now ready to use your newly flavoured luncheon meat.

There are lots of other ways to enhance your luncheon meat but the last one that I am going to cover is making your own homemade luncheon meat!! That's right you can make your own and colour and flavour it to suit you so the ingredients you are going to need are...

Blender
Aluminium Foil
2lb Ground Pork (or Beef)
180ml Water
1/4tsp Garlic Salt
1tsp Sea Salt
1tbsp Mustard Seed
10ml Flavouring (optional)
2tsp Powdered Colouring (optional)

Pour the water into the blender and add the 10ml of Flavouring and the 2tsp Powdered Colouring, put the lid on and quickly flick the switch to blend them together. Now put everything else in and switch on until well blended. Tip out onto the aluminium foil and shape into a rectangle. Fold the aluminium foil around your luncheon meat make sure it's wrapped well and put in the fridge for 24 hours. After 24 hours preheat oven to 170c, remove foil and place your luncheon meat on a raised rack in a baking (or old grill pan) and cook for 2 hours. Remove from oven and allow to cool, you now have your very own homemade luncheon meat.

HOME MADE BREADS

Another fantastic bait for fishing is bread, now you can obviously go to the supermarket and buy yourself a loaf of bread very easily but I thought I would give you a couple of recipes for making your own. As the recipes are you can happily eat these breads yourself as they are meant for humans but if you want to add 15-20ml of your favourite fishing flavouring you can customise your own flavoured bread.

CORN BREAD
250ml warm water
2 tsp active yeast granules
¼ tsp sugar
7oz plain flour
3.5oz fine ground cornmeal
pinch salt
1oz liquidized Sweetcorn
Place warm water in large bowl, add yeast and sugar and set aside. When yeast starts to bubble, add cornmeal, sweetcorn, flour and salt. Mix with one hand until dough is sticky. Transfer dough to well-greased or lined loaf tin. Cover with damp tea towel and leave to rise for about an hour. Bake the loaf in a preheated oven at 180 C for 30-35 mins.

SODA BREAD
6oz Self Raising Flour
6oz Plain Flour
Half tsp salt
Half tsp Bicarbonate of Soda
Half Pint of Buttermilk.

Now put the self raising flour, plain flour, salt and bicarbonate of soda into a mixing bowl and give a good stir to mix together. Create a well/dip in the middle of the mixture, pour the buttermilk into the well mixing quickly with a fork until you have a soft dough. Tip out onto a clean floured surface and knead for 5 minutes, form a round ball and place onto a lightly floured baking tray/sheet. Flatten slightly and cut a cross in the top. Place in the middle shelf of a preheated oven at 200 C and bake for 30 minutes. (Should sound hollow if you tap it) Place onto a wire rack and allow to cool.

HOME MADE FLOATER CAKE

I absolutely love surface fishing and one of my favourite baits for surface fishing is actually a golden oldie from the 80's or even older and that is floater cake. You will need the following ingredients in order to make floater cake and the advantage over normal bread is that it is slightly more robust and withstands the attention of smaller fish very nicely, you can also customise this more easily than bread to your own choice of flavouring.

2oz Semolina
2oz Cornmeal
1oz Milk Powder
1oz Ground Rice
6 Large Eggs
20ml Flavouring
10ml Glycerine
2tsp Baking Powder

2tsp Brewer's Yeast
1tsp Table Salt

To make Floater Cake place all your dry ingredients into an air tight bag, seal and give a good shake to mix everything up then set to one side. Now in a large mixing bowl whisk the 6 eggs until they are well mixed with lots of air bubbles. Now add your Flavouring and the Glycerine and give a good stir.

Next you want to slowly mix in the dry ingredients and you should end up with something similar to pancake mix. Grease a square cake tin with cooking oil and then pour the mixture in. Now place the tin into a preheated oven at 150degrees Celsius (Gas Mark 2) and cook for between 45 and 60 minutes.

Remove from oven and tip the floater cake out onto a cooling rack or plate and allow to cool. Once cut open you should have something similar to a nice moist sponge cake if all has gone well. (Although it will be firmer than sponge cake.)

If you want to try this with any of your 6 egg boilie mixes just double your eggs to 12 and then follow the instructions as above remembering to make your mix like a thick pancake mixture

HOW TO FLAVOUR SWEETCORN

One of the most popular and successfully used bait is sweetcorn, now I hear you asking what could I possibly be doing including sweetcorn in a book on making your own baits after all sweetcorn is cheap and ready to use straight out of the tin right? Well, the answer is yes... But, what if where your fishing has seen that much sweetcorn going in that the fish now associate it with danger and won't touch it?

The answer is home made flavoured and coloured sweetcorn the combinations of flavours and colours is almost endless and may just give you that extra edge that you need in order to catch those wary fish. Making your home made coloured and flavoured sweetcorn is actually very easy just follow these simple instructions and you will soon be experimenting with your own concoctions. If you want to add an essential oil into your mix do not use more than 3 drops or you will find it very overpowering.

You will need...

2 x 325g Tins of Sweetcorn drained and rinsed
1 Pint Cold Water
5ml Flavouring
10ml Glycerine
2tsp Powdered Colouring (or 28ml Liquid Food Colouring)
2 Pint Plastic Airtight Storage Jar

To make your Coloured and Flavoured Sweetcorn, pour the pint of water into the storage jar, now add the glycerine and flavouring and give a quick stir to mix. Now add your colouring and keep stirring until it is well and truly mixed with the water, tip your sweetcorn in and tighten the lid. Put it in the fridge for 12 hours, giving it a gentle shake after 6 hours. Finally take out and drain the excess liquid off and you are ready to use.

I will now give you a few flavour and colour combinations that have worked well for me but please don't forget the idea of making your own and feel free to experiment. One last tip, if you want a yellow bait still use yellow colouring as the flavouring may discolour normal yellow sweetcorn.

Strawberry Flavour and Red Colour
Aniseed Flavour and Red Colour
Scopex Flavour and Yellow Colour
Banana Flavour and Yellow Colour
Tutti Frutti Flavour and Orange Colour
Plum & Black Pepper Flavour and Purple Colour

Another great use for sweetcorn is Creamed Corn, not only does this make a delicious meal for you but it can be used as a Spod/Spomb mix and is also a great way if enhancing your hook bait. The following is a recipe that I have used for years and not only tastes delicious it has caught me lots of fish as well...

CREAMED CORN RECIPE
32oz Frozen Sweetcorn
4oz Butter

8oz Whipping Cream
1tbsp Sugar
Salt & Pepper (qty is how you like it)

Melt the Butter in a large Saucepan over a medium heat. Once the Butter has melted add the Frozen Sweetcorn stirring constantly for a couple of minutes until the Sweetcorn has thawed.

Now add the Whipping Cream, Sugar and Salt & Pepper and cook over a medium heat stirring constantly (or the Cream will burn) for 10-15 minutes until the sauce is nice and thick.

HOW TO PREPARE HEMP

Hemp seed has been a favourite bait of Anglers for many years due to its oil and flavour content that gets almost every species of fish feeding although it is traditionally used as a Roach bait. If you only want a small amount the easiest way to prepare it is to get a flask (the kind you keep coffee or tea in) then fill it half way with hemp seed.

If you would like to add some flavouring such as chilli add 1-2tsp now, fill the flask up with hot water put the lid on and either leave overnight or 12 hours whichever you prefer. Perfectly cooked hemp seed every time and the method I use most often.

Now the first thing I would advise doing is getting a large pot just for preparing your hemp seed as it will make a mess of the pot after so many times of use. For every pint of hemp seed you are preparing you will need to put in 3 pints of water then allow to it to soak for 24 hours.

Once you have soaked your hemp seed, add one teaspoon of salt per pint of hemp seed you have soaked. Bring the mix to a boil stirring occasionally and then reduce the heat and simmer for about half an hour.

If you are going to be using the hemp seed for Roach fishing this is the stage you need to be careful with (not quite so much if your just using it as feed), what you are trying to get is a few of the grains fully open, a few of them still closed and the majority of them just starting to split.

As soon as the hemp seed is cooked remove it from the heat and drain of the liquid (keeping the liquid as it is full of goodness and great for mixing into your groundbait). Give the hemp a rinse under some cold water then store in airtight bags in the freezer about a pint in each bag for ease of use.

Okay so you have your hemp ready to try for some nice Roach the best way to fish hemp on the hook is by taking A piece that has the white grain only just showing and gently pushing a size 18 hook into the split.

You must make sure that the shank of the hook is on the same side of the hemp seed as the blunt bit that attaches to the plant rather than the pointy end, all you do then is push the up and into the blunt end until you feel a gentle click, this will grip the hook nicely and keep it on the hook during casting. Make sure that the point of the hook is still showing or you will never connect with a bite. As for feeding your swim with hemp for Roach the best approach to get them fighting for your bait is to feed little and often feeding 15-20 grains of hemp seed every 5 minutes or after every cast if you are getting bites quicker than that.

So that is how to prepare hemp seed but I also have a few tips and tricks for you to try out...

1.. You can flavour your hemp seed with either 5ml flavouring per pint of hemp seed or 2tsp of dry flavouring per pint of hemp seed. Carp love hot hemp so you could add 2tsp of chilli flakes or if you want to target Barbel you could add 2tsp of garlic flakes. Add the flavouring to your dry hemp seed before you add the water.

2.. If you want to target bigger fish with hemp seed, create yourself a basic paste using, 2oz Semolina, 2oz Soya Flour, 1oz Milk Powder and 20g of Hemp Protein Powder. Slowly add this to 50ml of the water you boiled the hemp in until you have the texture of paste you like. Now wrap this around a hair rigged paste coil then take a handful of your prepared hemp and press it into the paste until no more will stick. Can be a bit of a pain to do but produces some nice fish.

3.. If your hemp seed dries out it will float so always keep it in the water you cooked it in when fishing. (or lake water if you are using hemp you froze)

HOME MADE BIODEGRADABLE PLASTIC

Okay so in this guide I will show you how to make a fairly tough plastic gel bait that will leak flavourings fairly quickly but will dissolve in water if it happens to come off. Fantastic for topping boilies, hair rigging on its own or even cut down to use as hook baits.

You will need the following ingredients;

6g Glycerine
22g Gelatin (unflavoured from the supermarket)
114ml Hot Water
1g Powdered Food Colouring (Mainline are brilliant)
2ml Flavouring (Nutrabaits are very good for this due to their strength)
5ml Additive (such as CSL or GLM powder)
3g Potassium Sorbate Granules

Now add all of the ingredients together into a small saucepan stirring constantly until there are no clumps of mixture left.

Put the saucepan on cooker/stove and heat until it starts to "froth". Make sure you stir the mixture constantly while you are waiting for it to "froth". As soon as your mixture froths remove it from the heat and keep stirring. With a spoon scoop out the excess froth making sure there are no lumps or clumps in the mixture.

Pour the hot liquid directly into your molds you can use whatever molds you like but I use a lot of Fritz Germany molds as they are ideal for using to create the shapes you want. Set the mold aside to dry, this will happen quicker if you place it in the fridge. Once dry, remove your bait from the molds and trim any excess off.

If you want a harder bait use less Glycerine, if you want your bait softer use more Glycerine. It may take several days for your bait to dry completely depending on humidity and temperature of where you're drying it so please be patient as you will end up with a safe and biodegradable alternative plastic bait.

HOME MADE GLUGS & DIPS

A lot of people have their own Favourite flavours in boilies but trying to get a matching glug can at times be almost impossible. I would like to show you how to make your own flavoured glug that will last for up to 1 month per batch you make. It is actually really easy to make your own and can be great fun but if you're using a smelly flavour make sure you make it in a well ventilated area or your other half may not be very impressed.

The ingredients you will need are:

200ml water
Sugar (use the same measuring cup you used for the water and measure out an equal amount of sugar, i.e. 200ml)

5ml Flavouring (if you use a Squirt such as Dynamite use 10ml)
10ml Food Colouring (optional)

To make your glug

Firstly put your water in a small saucepan and bring to the boil, now reduce heat to medium and add your flavouring and colouring and give a good stir. Now add your sugar and stir constantly until all the sugar is dissolved, allow to simmer, stirring occasionally until liquid is reduced by just under half. Remove from heat and allow to cool.

Pour into a glug pot or other airtight container and store in fridge. This will keep for up to a month.

One quick note to think about is that the liquid will thicken considerably as it cools so don't overcook it, you can always heat it again if it hasn't thickened enough.

This next home made item isn't bait as such, there has been a big trend recently to use a thick "Gooey" dip to coat your hook bait in creating an extra attractant for you to try and entice those wary fish. Now this may be a new trend but it is certainly not a new idea and this dip has been around in many forms for years, admittedly there are companies that have put new additives and attractants in making them even more enticing to the fish but you can make your own basic version quite easily and you can make it as thick or runny as you want it.

To make your Goo you are going to need the following ingredients...

Glycerine
Clear Honey
Flavouring
Food Colouring or Powdered Colouring

Firstly put your required amount of Glycerine in a mixing bowl, now add your colouring and flavouring (colouring can be adjusted as you go). Finally slowly add the Clear Honey stirring each addition in thoroughly until you have the required thickness of goo that you want.

Pour your goo into a container with an air tight lid such as a glug pot and store in a cool dry place, you may find a bit of separation occurs during storage just give the container a good shake before you use it.

HOME MADE CHEESE PASTE

One of my favourite all time baits is cheese paste when I'm making the paste the following recipe is what I use and have caught some very good fish on it, please feel free to look around and make other recipes but for now have a go at this one and see if it is right for you.

Right then you will need the following items:

340g Frozen Ready Rolled Shortcrust Pastry
160g Extra Mature Cheddar
160g Stilton or other Blue Cheese

To make your paste grate the extra mature cheddar cheese, then crumble the blue cheese and mix the two cheeses together. (you can alter the mix depending on what cheeses you have for example on the batch pictured I didn't have any blue cheese so I used 320g grated cheddar and 1tsp of paprika)

Spread the cheese mix over the pastry.

Fold over the pastry edges so that the cheese is trapped inside. Then gently knead the mixture until everything is all mixed together, keep kneading it until there are no lumps left. You should know when it is ready because the mixture will be roughly the same.

Now put the mixture into freezer bags and freeze until ready to use. This cheese mix is ideal for Chub but other fish will take it.

If you would like to tweak this recipe, you can always add a tsp of chilli powder, paprika, garlic powder or other spice, flavouring. You could even add some food colouring to alter the colour of your paste.

HOME MADE VANILLA BUTTERSCOTCH

A lot of people have asked me if I know how to make Scopex flavouring, the simple answer to this is no. However, I do know how to make a vanilla butterscotch sauce which smells very similar to Scopex whether it is similar or not I have no idea but it makes a great bait additive or glug.

To make the sauce you will need;

2oz Butter
3oz Soft Brown Sugar
1oz Golden Syrup
118ml Double Cream
8ml Vanilla Extract
1tsp Table Salt

To start making your sauce melt the butter in a medium sized saucepan on a medium heat (gas mark 3). As soon as the butter has melted add the sugar, golden syrup, double cream and the table salt.

Whisk it all together until it is mixed together nicely and bring the mix to a slow boil. Simmer for five minutes giving the mix a gentle stir occasionally.

Remove the saucepan from the heat and then stir in the vanilla extract and pour into a heat resistant container that have airtight lids but leave the lids off until the mixture has cooled completely.

You can add 10g of potassium sorbate to this mixture at the beginning if you wish to make the mix shelf life however, it will last for around 7-10 days in the fridge without preservative.

The next ingredient I would like to show you is corn steep liquor. I use Corn Steep Liquor (CSL) in a lot of my bait recipes as it is a mixture of soluble protein, amino acids, carbohydrates, organic acids, vitamins, and minerals which are easily metabolized and used by various species as sources of energy.

HOW TO PREPARE TIGER NUTS

Tiger nuts are another of my favourite baits and there are two ways to prepare them, I'm not going to say which is the best as there are equal amounts of people on each side who will argue that one method is better than the other.

The first method is very easy, place one kilogram of unprepared tiger nuts in a large saucepan. Pour boiling water over them so that they are covered by around two inches of water and leave them to soak for twenty four hours.

Pour the tiger nuts into a strainer after the twenty four hours allowing all the water to drain off. Then put them back in the saucepan, cover with fresh water and boil for thirty minutes. The tiger nuts are now ready, either use within twenty four hours or freeze until you want to use. With this method you can also allow them to dry out and then grind them down in a blender to a fine powder to create your own tiger nut flour.

The second method is slightly more complicated and will require a separate shed, garage or out house if you want to stay in your other half's good books.

This time place your one kilogram of tiger nuts into a bucket with a sealable airtight lid, add two hundred and fifty grams of granulated sugar (the stuff you put in your tea or coffee) and if you want to flavour them add 10ml of your chosen flavouring. If you want to colour them add ten grams of powdered food colouring now. (I like to add ten grams of Haith's Robin Gold)

This next bit is the same as the first method in that you want to cover the tiger nuts with boiling water until they are covered by around two inches of water. This time however, give them a good stir to mix in the sugar and any flavouring, colouring, etc that you have included. Seal the lid and place them in a nice dry warm place for four days.

After four days open the lid, give them a good stir and transfer them to a large metal saucepan and boil for thirty minutes. Once you have boiled them, pour them back into the sealable container and leave for another four days to continue fermenting. If you have done it correctly you will find that you have a sticky gooey substance all around your tiger nuts.

THE SIMPLE MAGGOT

Right then the next thing I would like to cover in this book is good old fashioned maggots. You might think how can you include maggots in a home made baits book well there are lots of things you can do with maggots the first and most simple of which is to flavour them, this can be done very simply by riddling them through a fine sieve to remove any maize flour which the shop usually put them in to stop them sweating then simply add around 5tsp of powdered flavouring such as Turmeric, Chilli or Garlic put the lid on, give them a quick shake and leave them overnight to take on the flavouring and hey presto maggots different to how others are using them.

The next way to use maggots (and some say the way to single out bigger fish) is to fish with dead maggots. There are a couple of ways to do this and as usual there are arguments on both sides as to which is the best way to do this.

Boiling them is the first method but be careful with this method as you can remove the colour from them if you're not careful, the best way to do this is to take your required amount of maggots and just cover them with cold water. Now slowly pour boiling water over them stirring constantly, once you are happy that they have all stop moving drain off the water and if you want add a bit of flavouring to them.

The second method of killing the maggots is much simpler, simply put the maggots into an air tight freezer bag squeeze gently till all the air is removed and freeze for a minimum of 48hours to ensure they are all dead.

A little trick with the freezing method is to add some flavouring (liquid or powder) into the freezer bag then put the maggots in and give them a gentle shake/rub to get them all covered in the flavouring, then you freeze as before but when you take them out of the freezer they will suck in the flavouring as they start to defrost. You may have heard of krilled maggots, well the freezer method I have just described is how to make them.

If you would like to give your hook baits that extra bit of attraction you can "gas" some of your maggots with flavouring. This is an old trick that has been around for years, basically you need an airtight container an old film canister is the ideal size but if you can't get one of those any airtight container of similar size will do.

Fill your airtight container with maggots but so that you can still fit the lid on, once you have done this put 2-3ml of flavouring and then put the lid on, leave for 24 hours and your maggots will be dead, full of flavouring and still look like normal maggots (or at least better than frozen dead maggots).

Okay so you have dead maggots, now what? Well there are a couple of ways I fish these, the first is with a cage feeder and I simply plug one end of the feeder with some groundbait then dead maggots and another plug of groundbait at the other end (I use Bait-Tech's Special 'G' Green but you can use which ever you are using). To this I use an 18" hook length with a size 12 Gardner Mugga Hook and two or three dead maggots on the hook. The second method is to use them float fished with the lift method with a size 16 hook and either a single or double maggot then put in small golf ball size balls of groundbait with dead maggots and hemp in them.

Lastly I make Maggot boilies, to make these you will need to liquidize the maggots (which might make you unpopular with the other half). Now to make these follow the boilie making instructions mentioned earlier in my book but use the following recipe...

pg. 106

MAGGOT BOILIES
4oz Semolina
4oz Soya Flour
4oz Milk Powder
2oz Cornmeal
1oz Fishmeal
1oz Wheatgerm
3 Large Eggs
Quarter Pint Liquidized Maggots
1tsp Brewer's Yeast
1tsp Betaine
1tsp Sea Salt

FRITZ GERMANY INSTRUCTIONS & RECIPES

I would like to include in my book a guide on how to use and make the most of a fantastic new product that I have found which allows you to make your own artificial lures, boilies and maize and this product is from a company called Fritz Germany.

To make your mixture use the following as a guideline on your required ingredients

Boilies/Lures:
100ml Warm Water (temperature must not be higher than 90°C)
50ml Gel Powder
15ml Pike Pro Mackerel Oil (optional)
1tsp Colouring (optional)
5ml Flavouring
1tsp Glitter (optional)

Maize (Sweetcorn):
20ml Warm Water (temperature must not be higher than 90°C)
10ml Gel Powder
5ml Groundbait/Powder (optional)
2.5ml Flavouring
1ml Colouring (optional)

Place your Gel Powder into a mixing bowl and then add the Water stirring constantly until the Gel Powder has dissolved.. Now add your Betaine +, Glitter and Particles/Groundbait, Flavouring, Colouring.

Now fill the provided syringe with mixture and inject into the moulds you are using. Once you have done this place the moulds into a cool area to harden, preferably a fridge.

As to how long they will last this varies greatly on the flavourings, particles, etc that you put into the mixture, if you store in the fridge in an air tight container you should be good for at least 2-3 weeks (although this is not a guarantee). You can if you wish to freeze the finished baits and then defrost them when ready to use with no apparent problem.

With the boilie moulds you can place a small cork ball in to make them buoyant or put half a ready made boilie in the bottom and fill with your flavoured gel. There are so many variations and options it's such a fantastic product. If you are struggling to find this new product in your local tackle shop you can get the whole range from...

http://www.mh-tackle.com/shop.html

RECIPES

The following are a range of recipes that I have tried successfully using these products, simply use the quantities mentioned above depending on whether you want boilies, lures or maize.

BAIT-TECH SWEET COCONUT
Warm Water
Gel Powder
Bait-Tech Special "G" Gold
Bait-Tech Sweet Coconut Liquid Squirt

SQUID & ORANGE
Warm Water
Gel Powder
Squid Meal
Blue Colouring
4 Drops Orange Essential Oil

PEAR & REDBULL
Warm Redbull (instead of Water)
Gel Powder
Milk Powder
Pear Flavouring

PLUM & BLACK PEPPER
Warm Water
Gel Powder
Plum Flavouring
2 Drops Black Peppercorn Oil
Betaine

Blue Glitter

CHILLI & CHOCOLATE
Warm Water
Gel Powder
Hot Chocolate Powder
Chilli Powder

STRAWBERRY
Warm Water
Gel Powder
Strawberry Milkshake Powder
Red Colouring
Strawberry Flavouring
Silver Glitter

ANISEED
Warm Water
Gel Powder
Red Colouring
Aniseed Flavouring
Silver Glitter

HONEY & SCOPEX
Warm Water
Gel Powder
Clear Honey
Scopex Flavouring

OYSTER
Warm Water
Gel Powder
Fishmeal
Oyster Sauce (from Supermarket)

SWEETCORN
Warm Water
Gel Powder
Corn Meal
Corn Steep Liquor
Yellow Food Colouring

CHOCOLATE ORANGE
Warm Water
Gel Powder
Hot Chocolate Powder
2 drops Orange Essential Oil

CHILLI & WHITE CHOCOLATE
Warm Water
Gel Powder
Chilli Powder
White Chocolate Flavouring

SWEET HALIBUT
Warm Water
Gel Powder
Liquid Pellet
Sweetener
Black Food Colouring

HOME MADE BOTTLE CAP LURES

Bottle cap lures are not a new concept they have been used in America and other countries for many years and have caught thousands of fish. These are now starting to make an appearance in the UK and the basic one is extremely easy to make, I thought I would show you how to make a couple of different types of bottle cap lures and I hope you enjoy making your own.

Okay so the first thing you will need is bottle caps, you should be able to get a quantity of these for free by going into your local pubs and asking them if they will save you some bottle caps. These ones in the photo were collected by my wife in a matter of twenty minutes from three different pubs, make sure you wash and dry the bottle caps to get rid of any liquid residue. Besides your bottle caps you will also need either a small screwdriver or a nine inch nail, a block of wood and a hammer.

Now that you have your bottle caps you will also need two split rings, a size 6 treble hook, two AAA shots and a size 8 swivel for each lure that you want to make. Take one of your bottle caps and place it on the block of wood then place the screwdriver/nail on the edge of the bottle cap and hit the screwdriver/nail with the hammer to make a small hole.

Turn the bottle cap round so that the hole is in the twelve o'clock position and then repeat the process creating another hole in the six o'clock position. Slide a split ring through each hole and then holding the bottle cap at three and nine o'clock positions squeeze it until it starts to fold over.

Once the bottle cap is nearly folded over so that the two sides touch each other, stop and put two AAA shots inside now continue to squeeze closed until the shots can't fall out when you let go. Finally slide the size 6 treble on to one split ring and the size 8 swivel on to the other and there you have your first basic bottle cap lure. If you don't want to make your own lures I found a page on Facebook where a student from Sparsholt College is selling these fantastic little lures. (www.facebook.com/bottlecaplures). This type of lure is ideal for Perch although it will catch other species too.

The next type of lure is a little bit more complicated to make and is a spinner style lure. In order to make this lure you will need; a bottle cap (prepared with holes the same as before), a size 6 treble hook, an 8 inch piece of strong but flexible wire such as fuse wire, a spinner blade with matching clevis, a split ring and finally a brass lure body. Personally I just buy these lure kits from America http://www.mudhole.com/Lure-Building/Lure-Building-Kits/French-Blade-Spinner-Kit it works out just over £1 per spinner including postage (but doesn't include bottle cap or split ring) and includes enough parts to make 20 lures. You will need two pairs of long nose pliers to bend your wire.

The first thing to do is to make an eye in the wire. This is actually fairly easy, place one end of your wire sideways in the end of one of the pairs of pliers with around one inch of wire sticking out one end. Carefully fold the one inch of wire around the end of the pliers until it passes the long piece of wire on the other side. Remove the wire from the pliers and you should have a loop in the wire if you've done it right.

Place the loop inside the pliers and clamp down tight, wrap the remainder of the one inch of wire around the larger stem of the wire (you may find it easier to use the other pair of pliers to do this). If done correctly you should end up with an eye in one end of the wire like the one shown in the photo.

Now you are basically going to slide the components on like you would if you were making a kebab. The spinner blade must be the first thing put on or the lure won't work correctly and to do this slide the blade on to the clevis (the little half circle piece of metal with holes at either end) then slide the clevis on to the wire making sure the top of the spinner is towards the eye. Next slide on the brass lure body followed by your bottle cap. You have choice here you can either leave the bottle cap as is or you can carefully bend it as before.

Lastly repeat the process of creating an eye but at the other end. Slide a split ring onto this eye and then slide the size 6 treble hook on and you now have a completed bottle cap spinning lure. You can add beads, propellers and more if you want the options are wide and varied.. This lure is ideal for Pike and Zander although as before it will catch other species too.

The final lure that I would like to show you how to make is the bottle cap cork popper and this you will need around 6-8" of wire, a wine bottle cork, a bottle cap, a few beads with a hole through them, a split ring and a size 6 treble hook.

To make the cork popper place the bottle cap on the block of wood face down and pierce the bottle cap through the middle. Next you want to make an eye in one end of the wire as you did for the previous lures, then slide the bottle cap on with the coloured part facing away from the eye.

Okay this bit is the hardest part of making this lure and that is you need to push the wire through the centre of the wine cork. Once you have done this (hopefully without stabbing yourself when it goes through the other end) thread a few beads on and the create another eye at the end of the wire making it as tight to the beads as possible. Slide the split ring onto the eye nearest the beads and finally slide the treble hook on to the split ring. You now have a surface cork popper.

USEFUL LINKS

I regularly get asked if I have any useful links for fishing related websites so I thought I would include a few here for you on this page. Links are working at time of publishing this book;

www.anglingdirect.co.uk (suppliers of the CentrePin featured on the front cover)

www.a1coffee.co.uk

www.crazy4flavour.co.uk

www.internationalegg.co.uk

www.lulu.com/spotlight/AnthonyWood

www.jerkbaitmania.co.uk (supplier of the Pike Fly on the front cover)

www.avena.co.uk

www.mh-tackle.com/shop.hmtl

www.bait-tech.com

www.haiths.com/supercatch

www.ccmoore.com

www.homemadeboilies.wordpress.com

RECOMMENDED COMPANIES

Now something I haven't done before is go into detail about a couple of the companies I use as I feel they need a mention in my book and a bit about why they get a mention.

The first company I would like to mention is Bait-Tech; I am part of their Street Squad due to them being a company that produces good quality products with a great company moral. I use their Oils, Corn Steep Liquors, groundbaits and method mixes in my boilie recipes and other bait recipes. Find out more at **www.bait-tech.com**

The second company is Haith's™ who have produced bait additives and ingredients to a high standard since the 1950's when they first started producing their famous Robin Red™ additive. Since then they have gone on to produce lots of other great additives and ingredients which I regularly use including their new Robin Green™, Robin Gold™, Robin Orange™, SuperRed™ and CLO™. You can find out more at **www.haiths.com/supercatch**

Finally the third company I would like to mention is Fritz Germany, this is a company that has been around for a while but is trying to get a footing in the UK Tackle market. I am a UK Consultant for them and they produce a range of silicon moulds for various lures, boilies and more. My favourite is the Sweetcorn (mais) mould kits although I am starting to get into my

predator fishing and am producing some good lures using the FB009 lure mould.
www.mh-tackle.com

WHERE I TEST MY BOILIES

I would finally like to give a description of my local fishery who have very kindly let me test all my baits on and have put up with me on a regular basis, so if you are ever in the Nuneaton area please pop in and give them a try. They also do a mean sausage and egg sandwich.

Barn Fishery, Atherstone Road, Hartshill, Nuneaton, CV10 0JB.
There are two lakes on this fishery. Which are available on a day ticket, which will be collected from your peg.

The first lake is a mature lake with an island in the middle, and a good selection of pegs available to the right as you are looking from the car park there are four pegs set up in a canal fashion and you will regularly see the bubbles of the larger fish patrolling the bottom. On sunny days you will be able to catch the odd wary Carp on a bit of floating bread. There are a lot of methods that work on this lake in particular luncheon meat is a good catcher, with maggots catching you a good selection of Roach (2lb+) and Rudd (1 ½ lb+). Also in the lake are a selection of Common and Mirror Carp up to and in excess of 26lb, Tench to over 8lb, Bream to 7lb+, Perch to 3lb+ and the odd surprise Ghost Carp and

Chub. This lake also has pegs that are suitable for the less abled angler.

The second lake is a work in progress and is being dug out into a match lake, even with work going on anglers are catching good bags of small Carp, Roach, and the popular F1's, with an occasional appearance of a larger Carp of which there a few up to approx 10lb.

There are no long walks and with car parking very close to the pegs, fishing these lakes is ideal for any capability, including novice and is in a good family setting with space for picnics.

Young anglers are welcome, however all children under 16 must be accompanied by an adult at all times.

Prices: (prices correct at time of publishing)

24 Hours   £17

Half Day   £4

FullDay   £6

Evening   £3

Directions:

From the B4111 Atherstone to Nuneaton Road when approaching the bend with a railway bridge above take the Hartshill turn. Pass 3 houses on the right after the 3rd house, next to the bus stop, take the track road (marked Kirby Glebe Fields). Continue to the very bottom of the track where you will find Barn Fishery.

FISH CAPTURE RECORD

Type of Fish Caught:_____

Date Fish Was Caught:_____

Weight of Fish:_____

Bait Fish Was Caught On:_____

_____

Location Fish Was Caught & Conditions:_____

_____

_____

pg. 121

FISH CAPTURE RECORD

Type of Fish Caught:_____

Date Fish Was Caught:_____

Weight of Fish:_____

Bait Fish Was Caught On:_____

Location Fish Was Caught & Conditions:_____

FISH CAPTURE RECORD

Type of Fish Caught:_____

Date Fish Was Caught:_____

Weight of Fish:_____

Bait Fish Was Caught On:_____

_____

Location Fish Was Caught & Conditions:_____

_____

_____

FISH CAPTURE RECORD

Type of Fish Caught:_____

Date Fish Was Caught:_____

Weight of Fish:_____

Bait Fish Was Caught On:_____

Location Fish Was Caught & Conditions:_____

FISH CAPTURE RECORD

Type of Fish Caught:_____

Date Fish Was Caught:_____

Weight of Fish:_____

Bait Fish Was Caught On:_____

_____

Location Fish Was Caught & Conditions:_____

_____

_____

FISH CAPTURE RECORD

Type of Fish Caught:_____

Date Fish Was Caught:_____

Weight of Fish:_____

Bait Fish Was Caught On:_____

Location Fish Was Caught & Conditions:_____

FISH CAPTURE RECORD

Type of Fish Caught:_____

Date Fish Was Caught:_____

Weight of Fish:_____

Bait Fish Was Caught On:_____

Location Fish Was Caught & Conditions:_____

FISH CAPTURE RECORD

Type of Fish Caught:_____

Date Fish Was Caught:_____

Weight of Fish:_____

Bait Fish Was Caught On:_____

_____

Location Fish Was Caught & Conditions:_____

_____

_____

A SELECTION OF BONUS MATERIAL

In the following pages I will post a few last minute bits and bobs of information that I think will be of use to you.

The first thing I would like to show you is a rough protein guide for some of the more common ingredients...

One Large (50ml) Chicken Egg
Protein is Roughly 13%

One Ounce (28g) of Semolina
Protein is Roughly 7%

One Ounce (28g) of Soya Flour
Protein is Roughly 26%

One Ounce (28g) of White Rice Flour
Protein is Roughly 3%

One Ounce (28g) of Full Fat Milk Powder
Protein is Roughly 15%

One Ounce (28g) of Wheatgerm
Protein is Roughly 13%

One Ounce (28g) of Whey Protein
Protein is Roughly 40%

Now an important thing to remember after reading about basic protein levels is that baits are not only about the protein levels but also about amino levels, solubility, digestibility, mineral content and more.

This is one of the reasons I try to stay away from the science side of things as much as possible when on my Facebook page or writing my books as it can get highly technical, very confusing and people will always argue that their way of doing something is right whereas there are actually lots of ways of making your own baits which is why I like to keep it as simple as possible and let you enhance or learn more as and when you feel ready to.

The biggest thing to remember is if you find a way that works, as long as it is safe for the fish then don't let anyone convince you that your way is wrong.

RAGI MUDDE

This is a very simple bait that has been used in India for years to catch the rather magnificent Mahsheer. You will need the following ingredients:
235g Ragi Flour (Millet Flour)
470ml Water
2tsp Salt

To make the Ragi Mudde place 235ml of water in a mixing bowl and slowly add the flour to the water stirring constantly, you should end up with a smooth thick paste. If you want to spice it up a bit you can add 2tsp spices, garlic or other flavouring making sure it is stirred into the paste well.

While you're making this paste bring the other 235ml of water and salt to a boil. When the water starts to boil add your paste, stir constantly with a wooden spoon until it all comes together.

Tip out onto a plate and carefully give it a knead for a minute, allow the dough to cool for about 5 minutes.

Next wet your hands with a little water, break the paste into two and roll each piece into a ball. The Ragi Mudde is now ready.

Pinch off pieces as and when you want to you use them, the Ragi Mudde is actually delicious to eat as well and tastes great served with a salsa of cucumber, onion, tomato and celery.

VANILLA CHAI MEAL

A lot of people have asked me if there is an alternative to vanilla meal which is a popular ingredient. I do have an alternative which is vanilla chai meal which is a good source of protein, antioxidants and more. It isn't necessarily cheap to make but is a fantastic ingredient for using in your bait making.

Ingredients...
16oz Soya Flour
8oz Semi Skimmed Milk Powder
8oz Coffee Mate
8oz Whey Powder
4oz Sweetener
20g Vanilla Powder (4tsp)
10g Ground Ginger (2tsp)

10g Ground Cinnamon (2tsp)
5g Ground Gloves (1tsp)
5g Ground Cardamom (1tsp)
5g Nutmeg (1tsp)
5g Allspice (1tsp)
2.5g White Pepper (half a tsp)

To make vanilla powder simply microwave whole vanilla pods in 10 second bursts until they are dried out then blitz them in a coffee grinder until they are powdered.

Place all of the ingredients into either a mixer or a large air tight bag and mix it all together thoroughly. Once it is all mixed together use at up to 4oz in a 16oz base mix.

You can also make a rather delicious tasting Vanilla Chai Tea by simply removing the Soya Flour and using the following recipe…

16oz White Sugar
8oz Semi Skimmed Milk Powder
8oz Coffee Mate
8oz Whey Powder
4oz Unsweetened Instant Tea
20g Vanilla Powder (4tsp)
10g Ground Ginger (2tsp)
10g Ground Cinnamon (2tsp)
5g Ground Gloves (1tsp)
5g Ground Cardamom (1tsp)
5g Nutmeg (1tsp)
5g Allspice (1tsp)
2.5g White Pepper (half a tsp)

To make your Vanilla Chai Tea add 40-60g of the above mix to 8fl.oz of either hot water or hot milk.

FRUIT FLAVOURED SYRUP

A fantastic way to create a glug for your fruit boilies is to create a fruit flavoured syrup. These syrups can also be used to flavour your deserts, coffee, ice cream, etc as well. You will need the following in order to make the syrup.

A Large Saucepan
A Heat Resistant Airtight Container
4oz Water
8oz Granulated Sugar
4oz Fruit Juice/Pulp
2oz Fruit Peeled and cut into pieces (must match the juice/pulp)
3g Potassium Sorbate
A fine mesh strainer.

To make your syrup is very easy, put the water in the large saucepan and bring to the boil. Once the water is boiling turn off the heat and add the potassium sorbate, fruit juice/pulp, fruit and sugar.

Keep stirring until all of the sugar has dissolved. Pour the mixture into the airtight container(s) and allow to cool.

You now have an option as to how long to leave the fruit pieces in the syrup. The longer you leave the fruit pieces in the stronger the flavouring will get.

Once you are ready to separate the fruit pieces from the syrup simply put your syrup through the fine mesh strainer and the fruit pieces will be left in the strainer giving you a nice smooth syrup.

THE FORGOTTEN ART OF FLOAT MAKING

I have written this bonus chapter in the hope that it will encourage a few more people to look at the way in which you fish and that it will open up your view on the world of angling. Today's society has become an "instant" world where you can walk in to a shop and within a couple of hours have everything you need to become an overnight angler, to many this is a satisfying and rewarding way to fish and there is nothing wrong with that but have you ever wondered why a float is a particular shape or imagined the feeling of watching a float you have made sail away under the water and indicate you have a fish?

It is these sorts of feelings and knowledge that I feel are being lost to our sport so I thought I would write a step by step guide showing you how easily and cheaply you can make your own floats. The first thing you will need to do is to get your raw materials for making floats, now there are lots of materials you can make floats from such as reed stems, bird feathers, porcupine quills, twigs and even a plain old straw but for this guide I am going to use balsa. Okay so what are you going to need? Here is the list you will need in order to follow this guide and make your own floats;

Balsa Wood diameter of your choice (you can get this either online or from most hobby/craft shops)
Black Permanent Marker Pen
Red Permanent Marker Pen
A Tub of Corrective Fluid
Wooden Kebab Skewers
A Sheet of Fine Sand Paper
A Sharp Craft Knife
Super Glue
A Modelling Paint Brush Small
A Size 8 Swivel
A Baiting Drill or Small Drill Bit
Waterproof Clear Varnish
Pair of Pliers
Junior Hacksaw

So you have all of your materials, the next thing to do is to cut a piece of Balsa between 4-5 inches long. Very carefully at one end drill a small hole in the middle about 2-3cm deep, next you need to cut the loop off one end of a swivel using the pliers. Hold the remaining loop of the swivel in the teeth of the pliers and gently force the barrel of the swivel into the hole in the balsa, don't worry that the hole is smaller it will expand as you push the swivel in. Once all of the barrel has been pushed into the balsa, gently pull it back out and squeeze a couple of drops of super glue into the hole then reinsert the barrel. Set this aside and allow to partly dry for about an hour.

Next you need to very carefully take small shavings off the piece of balsa at the end with the swivel in until the balsa is roughly blended in to the end of the swivel barrel. Now take your sandpaper and gently sand both ends into a nice smooth finish, carefully blowing any balsa dust away. The next stage is the fun one, you get to colour in your float I will show you how I colour mine but you can be creative with your designs and colours.

Firstly, take the correction fluid and apply it to the end of your float without the swivel covering about 4-5cm of the balsa. Allow this to dry for a couple of minutes, take your black marker pen and draw a circle around the balsa where the correction fluid meets the balsa. The easiest way to create this circle is to hold the pen still and rotate the float, once you have this circle carefully colour in all of your float below the circle right down to the swivel making sure it is all black with now gaps anywhere. Repeat the circle process with your red marker leaving a small ring of correction fluid between the black and your red circle, colour in the float from your red circle to the top of the float.

Leave everything to dry for 5 minutes then clamp the swivel loop between the plier's teeth again and carefully paint a layer of varnish on to your float. I have completed my floats like this before but to make it completely waterproof allow the varnish to dry for 24 hours then paint another layer of varnish on just to completely seal it, again allowing it to dry for 24 hours. You now have a completed float ready to use, test it in your sink by using a length of line and adding BB shots until it sits how you want it to, remember the amount of weight required or you will be constantly fiddling about on the bank getting it right.

This type of float is perfect for all types of float fishing on commercial lakes and if you need to create a distance float simply cut yourself a longer piece of Balsa at the beginning meaning that you will need more weights to set the float correctly which means you will get more distance on the cast. I have been using these for years and find that I can even get positive indications from 1oz Gudgeon.

The next type of float I want to show you how to make is a stick float these types of floats are ideal for River fishing and it is very easy to make. The first thing to do is take a wooden Kebab Skewer and from the pointed end measure roughly 6 inches, mark this up and then using the junior hacksaw, gently saw the skewer keeping the 6 inch section and discarding the rest (or keep it for making different floats). Now cut yourself roughly a 4 inch section of the balsa wood and using the baiting drill make a 2 inch hole thought the middle of one end of the balsa wood. Carefully push the spiked end of the skewer into the hole that you have made in the balsa wood until you have a tight fit, remove the skewer and squirt a small amount of super glue into the hole and reinsert the skewer set it to one side and leave to partly set for around an hour.

Carefully take your knife and trim the edges around the top of the float leaving a vague dome shape. At the other end of the float slowly take small shavings off the balsa wood until the ends are touching the skewer. Taking your sandpaper sand the top of the float into a nice dome shape, now this bit is slightly different than the other float in that you want to start sanding about half way down the balsa and being careful sand the lower half of the balsa so that it slopes into the skewer creating a tear drop type shape of the whole balsa section. Now you can colour your float in your chosen colours, I have again opted for my favourite pattern of red, black and white. Finally apply 1 or more coats of clear varnish to finish off your float.

If you are not familiar with this type of fishing you may have noticed that there is no swivel on this float, this because you need to get small circles of rubber called "float rubbers" you can pick up packs of around 50 assorted sizes for about £1-£2. You need to place a minimum of 2 float rubbers on the stem (the skewer) of your float although I always recommend using 3 incase one snaps. You want the float rubber to be tight fitting as you will run your line between the rubber and your float stem to hold it in place. To weight this float you will need to spread your weights out along the line between the end of your float and your hook, I won't go into shotting patterns of your weights as it all depends on how you want to fish, how fast the rivers flowing and lots of other factors but to work out approximately how much weight you need just run some line through the float stops and keep adding weights on until it sits nicely in your bath, sink or whichever water container you are using.

The final type of float I would like to show you how to make is a float that looks quite strange, these floats have a large body, a thin neck and finally a medium sized head. What is the reason for making a float of this shape? Well the answer is quite simple really the large body will be fully submerged under the water creating less drag in windy conditions, the thin neck allows water that is being affected by the wind to pass by fairly unblocked and the medium sized head allows you to see the indication of a bite a lot better.

So how do you make this strange looking float? Well it's actually very similar to the first float I showed you how to make, you start out by taking around 2inches of balsa wood but this time using the baiting drill put a hole in the middle of both ends one just deep enough to fit the swivel in (you can glue the swivel in to place now if you like) and the other around half an inch deep.

The next step is to take a piece of skewer around 2-3 inches long and glue it into the other end, now you will need a smaller piece of balsa wood for the head of the float and you will need to cut around half an inch of this. Very carefully drill about 2cm into one end of the balsa wood and glue it on to the end of the skewer, set aside and allow to partly dry again for an hour. Now using your knife carefully shave of the edges of the balsa wood to create a rounded effect, sand everything down so it's nice and smooth and colour and varnish like your previous floats.

Well hopefully by now you will have found a new aspect of fishing to enjoy and the relaxing concentration needed to make your floats will give you something else to do on those days where you just can't get out to go fishing. There is something magical about admiring a collection of floats that you have made yourself. Please take the time and give it a try you may just find that you have found yourself a new hobby.

Well that is my final book completed. I hope you have enjoyed reading it and have picked up some new tips and tricks to help you make your own baits. Good luck with your bait making and feel free to join me on my Facebook page...

http://www.facebook.com/HomeMadeBoilies

Lightning Source UK Ltd.
Milton Keynes UK
UKOW06f1801240615

254069UK00007B/319/P